TORRES STRAIT

NEW GUINEA

140°

Gove

GULF

OF

CARPENTARIA

Groote
Eylandt

CAPE

YORK

PENINSULA

Pascoe River

McIlwraith
Range

Coen

TABLELAND

Y

reek

Normanton

Mt Spurgeon *

ATHERTON
TABLELAND

Ravenshoe

Cardwell

5

9

Cooktown

Daintree River

Mossman
Cairns

Innisfail

Hinchinbrook I
Herbert River

Magnetic I

Townsville

SOUTH

CORAL   SEA

0   100   200   300   400   500

miles

PACIFIC

Mt Isa

Cloncurry

Hughenden

Prairie

Mackay

Percy I

ka
Tarlton
Ra

S

QUEENSLAND

Clermont

Rockhampton

20°

OCEAN

Birdsville

Diamantina River

Barcoo River

Dawson R

1

5

9

Mary
River

Gympie

Cooncherie

Creek

Cooper

Wilson R

Range

Cunnamulla

St George

DARLING

DOWNS

BRISBANE

160°

Stanthorpe

Murwillumbah

Richmond River

Clarence River

tta

re

Grey

Main Barrier Ra

Moree

Dorrigo

30°

MURRAY   RIVER

HUME
RES.

Tallangatta

Flinders Range

DARLING RIVER

Bourke

Manilla

LIVERPOOL
PLAINS

Hastings River

Carrieton

Broken
Hill

Wilcannia

Nyngan

NSW

0 10 20   50

miles

Mt Wills

*

ange

LIA

5

Morgan

MURRAY

Trentham
Cliffs

Lachlan River

Euston

Booligal

Murrumbidgee River

Port Stephens
Hunter River

Newcastle

Gosford Lake Macquarie

Jenolan

SYDNEY

Wollongong

Mt Hotham * RANGE

Timbertop *

EILDON
RES.

*

ADELAIDE

Swan Reach

Fromm's Landing

Lake Alexandrina

Lake Albert

RIVER

Deniliquin

Wombeyan

Lake Illawarra

CANBERRA

GREAT

Goulburn

Lake Mtn

Marysville

DIVIDING

RIVER

ncer Gulf

ULA

rool

Nhill

VIC.

Gilmore

Brindabella

AUSTN

ALPS

Montague I

MELBOURNE

9

Robe

Cooronq

Halls Gap

Grampians

* Mt Alexander

* Mt Cole

GIPPS

MELBOURNE

Tanjil Bren

Noojee

Thomson R

LAND

Mt Gambier

Stony
Rises

MELBOURNE

Orbost

DURDIDWARRAH
RES.

Dandenong
Ranges

GIPP

Latrobe River

Portland

Lady Julia Percy I

Port Fairy

Cape Otway

©

Wilsons Promontory

PORT
PHILLIP
BAY

S

Bass River

BASS   STRAIT

King I

Flinders I

Cape
Barren I

MORNINGTON
PENINSULA

WESTERN PORT
BAY

Phillip
Island

WILSONS
PROMONTORY

Smithton

Wynyard

Kelso

* Cradle Mt

Strahan

TASMANIA

HOBART

© BASS   STRAIT

140°   150°

A Guide to the
Native Mammals
of Australia

William David Lindsay Ride, M.A., D.Phil. (Oxon.), was awarded a Doctorate of Philosophy as a result of studies in the evolution and relationships of the Kangaroo family. Since 1957 he has been the Director of the Western Australian Museum and is the author of many scientific papers on Australian mammals. He is much involved in conservation, as a member of the Western Australian Wild Life Authority, of the Australian Academy of Science committees on national parks, reserves, and fauna and flora, and as a council member of the Australian Conservation Foundation. He is also an Honorary Reader in Zoology in the University of Western Australia, where he lectures on the evolution of vertebrate animals, and is the Australian Commissioner on the International Commission on Zoological Nomenclature.

Ella Fry studied at the East Sydney Technical College Art Department, now known as the National Art School. In 1938 she was awarded the Diploma A.S.T.C. and the Bronze Medal for Highest Honours. She has held exhibitions of paintings and drawings and for several years gave lectures on art for the Faculty of Education in the University of Western Australia and for Adult Education classes. She is a member of the Art Gallery Board in Western Australia, the first woman to be appointed, and is represented in the Queensland and Western Australian art galleries.

RIDE, W. D. L. **A guide to the native mammals of Australia.** with drawings by Ella Fry. Oxford, 1970. 249p il map bibl. 10.75. SBN 19-550252-3

CHOICE *JAN. '71*
Reference

The native monotremes, marsupials, rodents, bats, and carnivores of Australia are described and superbly illustrated by pen and ink drawings made from life by Ella Fry. Each genus, or group of related genera is descussed and the species are briefly characterized. Considerable attention is given to the ecology and habits of the animals, to the effect that man has had on their distribution, and to problems of conservation. There are chapters on the rare ones, the ones in danger, and on the purposes of conservation. Additional reading suggestions and references to the technical literature are included as appendices. This is the first comprehensive study of Australian mammals to appear in many years. It is an excellent introduction for the layman and beginning student. Strongly recommended.

A Guide to the
Native Mammals
of Australia

W. D. L. RIDE
The Western Australian Museum

With drawings by Ella Fry

MELBOURNE
OXFORD UNIVERSITY PRESS
LONDON WELLINGTON NEW YORK
1970

Oxford University Press, Ely House, London, W.1

GLASGOW NEW YORK TORONTO MELBOURNE WELLINGTON
CAPE TOWN SALISBURY IBADAN NAIROBI DAR ES SALAAM LUSAKA
ADDIS ABABA BOMBAY CALCUTTA MADRAS KARACHI LAHORE
DACCA KUALA LUMPUR SINGAPORE HONG KONG TOKYO

Oxford University Press, 7 Bowen Crescent, Melbourne

First published 1970

SBN 19 550252 3

Registered in Australia for transmission by post as a book
PRINTED IN AUSTRALIA BY HALSTEAD PRESS, SYDNEY

Contents

Illustrations

Marsupials

Rodents

Bats

Carnivores

Monotremes

Introduction

ONE HUNDRED years ago, when John Gould sat down to write his Introduction to the *Mammals of Australia*, he was mostly concerned with telling people about the discovery of the different kinds of mammals which occurred in the then newly occupied island continent—and in the discovery of which he had played such a conspicuous part. But, even in those early days, he found it necessary to speak of the problem which would be faced by the Australian naturalist in days ahead; so he wrote:

> .... Australia—a part of the world's surface still in maiden dress, but the charms of which will ere long be ruffled and their true character no longer seen! Those charms will not long survive the intrusion of the stockholder, the farmer, and the miner, each vying with the other to obliterate that which is so pleasing to every naturalist; and fortunate do I consider the circumstances which induced me to visit the country while so much of it remained in its primitive state.

Today we see Gould's vision of nature in an altered land. Land utilization in Australia is rapidly approaching its inevitable end—the day on which there is no land left uncommitted to man's use. On that day, the only virgin bush left will be that which has been specifically set aside to conserve it.

As that day comes nearer, we, the people who live in the land, must recognize, face, and plan for the great challenge of the future—the challenge to conserve our natural resources, while we use and enjoy them, be they soils, forests, or individual plants and animals.

This guide to the mammals of Australia, written in 1968, about the mammals of an altered and altering continent, is as much an introduction to the problems of conserving the mammals as it is a guide to the kinds of mammals themselves.

The illustrations

The book is constructed about the theme that the 228 species of Australian mammals listed within it can be readily recognized as members of one or other of 55 groups of species (see p. 36). To enable you to recognize the general characters of any group, it relies on drawings. The words of text then confirm the general characters and tell you something of the habits of the mammals, where they may be found, and the characters of the individual species.

To achieve this aim, we have chosen to use drawings rather than photographs because we believe that they are the best means of displaying mammals in groups of poses which are characteristic and revealing. At first sight, it may be thought strange that we have chosen to use monochrome and not colour. In making this choice, we have decided that the attractiveness which the use of colour would bring is a minor advantage as compared with the lower cost of monochrome to the user; but also because in the identification of groups colour is rarely necessary and may even hinder recognition, since it is most often characteristic of species while general form is the most important group feature. Colour often transgresses the groups but their physical characters define them.

In order to ensure that the poses presented in the drawings are characteristic, Ella Fry has kept as many as possible of the animals in captivity—some for many months— so that she could watch and become familiar with them before selecting postures for her drawings. This would not have been possible without the ready co-operation of the many friends from all over Australia who have made it their business to supply us with living animals. We are also indebted to the Western Australian Department of Fisheries and Fauna who, as a result of their interest in the project, have allowed us to keep the animals in captivity.

In the case of those species which we have not been able to keep captive in Perth (see p. 236 for details) the plates are based upon photographs. In most of such cases, each illustration is made up from a number of photographs so that detail most useful in identification is revealed.

In looking at the plates, the reader should bear in mind that each contains three kinds of factual information. First, there is information on the physical features of the animal and its poses. We believe that these are not only characteristic as to posture and to structure, but they are also as accurate in proportion as it has been possible for us to make them, because we intend them to be of precise zoological use as well as conveying the attraction which the animals have to the artist. Even when the illustrations are studies of living animals accuracy of proportion, and of posture in rapid motion, have usually been ensured through reference to numerous photographs which have been taken for comparison with the drawings. At times, too, additional confirmation has been sought from measurements.

Where more than one individual is included in a plate, the resulting group gives a second kind of information; but the reader must not be led astray by it. In these illustrations it has been our purpose to present several characteristic poses of each species within the space of a single plate. We have done this in order to give a pattern akin to that seen by an observer when an animal is going about its business. By this means the reader is helped to surmount the difficulty of identification which he will find in many animal books: a difficulty which is usually caused by the reader's inability to decide whether the differences that he sees between different illustrations are real, or are merely the result of animals being presented from dissimilar viewpoints, (or even, when animals are uniformly presented, as in some field guides, where critical characters may be seen better from different aspects). Unfortunately, this choice has meant that behavioural realism may have had to be sacrificed because many mammals are not social animals; in such cases mammals will seldom be found in close proximity in the wild unless mating, fighting, or threatening and submissive. Accordingly, the reader must not interpret these illustrations as groups depicting social behaviour unless the description in Appendix III (pp. 236-40) warrants it.

The third piece of information in each illustration is the environment which it pictures. Remember, in nature, very few of the subjects would be seen before sunset except in thickets, in holes, or under litter, but here they seem to be in full light. Many of the drawings have been made at night under subdued lighting; through them, we can see, through the artist's eyes, the animals moving around, unafraid, under the cover of darkness, but with the darkness removed. In the interest of presentation, the amount of material in the background has also been reduced, but Ella Fry has tried throughout to introduce some characteristic natural elements into each picture in order to convey what she feels about the habitat of her subject.

Acknowledgements

ANY BOOK written today on Australian mammals builds upon the work of its predecessors. Those reading this book who are familiar with other works published on our mammals in the past will notice such things as differences in the number of species that are recognized and even changes in statements about where animals occur. This is because our forerunners, through their works, revealed gaps in knowledge which have since been filled; but in their writings they assembled the information which provides the stimulus and foundation for our own.

In addition to the debt which we owe to those who have gone before us, we must also acknowledge what we owe to our contemporaries. This book differs from earlier works on Australian mammals because present-day mammalogists and field workers are discovering what is new and reinterpreting what was known before. In particular, this book owes much of its present form to our colleagues John Calaby, of the Division of Wildlife Research, C.S.I.R.O., and Jack Mahoney of the University of Sydney. The classification adopted here is based upon a new checklist of Australian mammals which Calaby, Mahoney and I have in preparation and, in presenting this book we are grateful to them for allowing us to incorporate so many of the results of this unpublished work. In addition they and John Kirsch also read and commented on the text and the user owes much to their constructive and penetrating criticisms.

This book is almost entirely illustrated from living animals. The extent of this would not have been possible without the very great assistance and skill of Harry Butler who, with the approval of Hobart Van Deusen, the Curator of the Archbold Collections at the American Museum of Natural History, made available to us numerous living mammals which he collected during his expeditions for the American Museum and the Western Australian Museum. Other living mammals were provided for us by a large number of people whose help is acknowledged below.

That we were allowed to keep the animals captive at all was due to the sympathy and interest in the project of Mr A. J. Fraser, the Director of Fisheries and Fauna of Western Australia, and to Mr H. Shugg, the Chief Warden of Fauna of Western Australia.

Numbers of people, both in their private capacities and in official positions, have helped us in preparing this work by obtaining animals for us, helping us to keep them, providing us with information, or checking what I have written. In addition my secretary, Mrs Lyn Wieck, bore the whole burden of preparing the manuscript, and my Research Assistant, Mrs Helen Henderson, helped in ways too many to define. Sections of the text were checked for accuracy by numbers of authorities and, in particular, I must acknowledge the debt which we owe to Miss Joan Dixon, Mr Peter Aitken, Mr Alan Bartholomai, Mr John McKean, Mr Basil Marlow and Mr Keith Dempster who, as Curators responsible for the State and Commonwealth Collections of Mammals, checked the records given in the chapter 'The Rare Ones'. Mr Norman Wakefield provided me with much useful information on animal distributions in Victoria which was not otherwise available to me. Those who helped in other ways

were: The Fisheries and Wildlife Department, Victoria; The Directors of the Austral-
ian Museums; the Zoological Gardens Board and the National Parks Board of
Western Australia; the Zoology Department and University of Western Australia;
The Animals and Birds Protection Board, Tasmania; The Fisheries and Fauna
Department, Western Australia; TVW Ltd, Perth; Miss J. Arnold, Mr and Mrs John
Bannister, Mr Alex Baynes, Mr and Mrs Neville Beeck, Mr R. A. Breeden, Mr S.
Breeden, Mr J. E. Carr, Dr and Mrs Peter Crowcroft, Mr Keith Dempster, Mr G. M.
Doak, Mr Athol Douglas, Dr E. H. M. Ealey, Dr John Flux, Mrs G. B. Frieze,
Mr E. Garratt, Dr Eric Guiler, Mr A. S. Hill, Mr and Mrs R. Hiller, Miss Judith
King, Dr Gordon Lyne, Mr Roy D. Mackay, Mr L. N. McKenna, Mr John McNally,
Professor A. R. Main, Mr Michael Morcombe, Mr T. K. Moriarty, Dr J. Nelson,
Mrs A. Neumann, Dr and Mrs E. Pianka, Mr K. Richards, Master Edwin Ride,
Mr J. H. Seebeck, Mr K. Sheils, Mr G. Selk, Mr and Mrs Peter Slater, Mr Tom
Spence, Mr B. Stagg, Mr John Tapper, Constable R. Taylor, Dr C. H. Tyndale-
Biscoe, Dr Donald Walker, Professor H. Waring, Mr R. M. Warneke, Mr John
White, Mr S. R. White, Dr B. R. Wilson, Mr J. Womersley, Mr J. T. Woods and
Dr Patricia Woolley.

The Board and Staff of the Western Australian Museum have a special place in
these acknowledgements because no comprehensive work of this nature can be
successfully completed today solely through the resources of the individual worker.
The field work, the collections, the specimen preparation, the documentary records,
the libraries and the professional consultation which underlie a modern work of
natural history require the backing of institutions which have for their fundamental
purposes the production of knowledge and the charter to make it available to men.
In fulfilling this purpose the author and the artist are only the architects and tools
whereby the end is achieved. Throughout the work that led to this book, Sir Thomas
Meagher and the Board of the Western Australian Museum have generously made
available the facilities which have made it possible.

Finally, that we commenced at all, persevered and concluded tells of the parts
played by Margaret and Mel and our debt to them is beyond words.

1

Australian mammals today

IF YOU LIVE in Melbourne, or Sydney, or even Perth, and you take a holiday in Alice Springs—and are lucky enough to see some native mammals—you will find that they are not the kinds (or more properly, species) with which you are familiar in the national parks around your own home. In a general sense they will not be completely different—but neither will they be identical. For example, you will still see kangaroos, but they will be the Reds or Plains Kangaroos not the Grey Kangaroos of the coastal eucalypt country.

'But of course', you will say, 'Alice Springs is not Melbourne—the country around is different.'

Country, or the environment of a mammal, is made up of a whole host of component parts. Of these, the factors which exert the greatest overall influences on mammals are climate and the nature of the soils, or other substrate, which control the plant cover upon which they depend to provide their food and shelter. This cover, its density, and the presence of other animals, all together affect the life of the individual mammal. One species of mammal requires one set of conditions, while another species will require a different set; some may be wide in their tolerance, and they can be found in several kinds of environment, while others are very narrow in their demands and can only exist where their precise needs are met. The science which describes these factors and their impact upon an animal or plant is called ecology, a word derived from the Greek words *oikos* meaning a house, and *logos*—discourse. When we talk about the requirements which a species may have, we call them its ecological requirements, and the man who studies them is an ecologist.

Before man came into Australia and broke up the broad sweep of plant distribution with his fires, his stock, and his clearing, the two things that determined whether any mammal could occur in a particular environment were the combination of its ecological requirements and its past history. Thus, although the ecological requirements of a species might be fulfilled in one place, yet if it evolved elsewhere and was prevented from getting to the suitable place by unsuitable terrain, then it will not be found in both places although the environments of both may seem identical. For example, the Greater Gliding Possum, which almost certainly evolved in the great eucalyptus forests of the south-eastern mountains, is unknown in the great eucalypt forests of the south-west. The explanation may be that conditions to the north of the head of the Australian Bight have never been suitable for them to be able to move across southern Australia from east to west.

Much about the distribution of modern Australian mammals, before man upset it, is revealed by specimens collected in the early days of Australian biological exploration before man drastically altered much of the country. These reveal general principles of distribution which it is important to try to visualize; from them, we can

1

infer distributions where today we can only see the remnants of them and, in trying to identify mammals, knowledge of these distributions enables us to say that such-and-such a species might well be expected to occur in one place, and not in another.

First of all, the major patterns of mammal distribution seem to be controlled by climate and in particular upon the time of the year during which the seasonal rain falls and the extent of this rainfall. Thus, in northern Australia, rain is monsoonal and falls in the summer. In the south-west and south-east, rain is controlled by the circumpolar weather systems and, as a result, the rainy season is the winter. In the centre, on the other hand, although most rain is summer rain, it is intermittent and irregular and, in addition, it is very low in quantity. As a result any one place only receives rain at very irregular intervals, and although it may rain very hard indeed when it comes, rain may only fall in significant quantity at intervals of several years.

These patterns of factors controlling mammalian distribution within Australia are shown in the endpaper map.

The East Coast:

If we first look at the east coast of Australia on the map, we can see that (unlike the west coast) there is a long, almost continuous, border of effective high rainfall. This is the result of the great chain of mountain ranges which extend from the Grampians and Dandenongs, in Victoria, and then northwards through the Australian Alps right up to Cape York Peninsula.

The effect of this range is not only to increase rainfall by precipitation from the moist Pacific air, but also to modify the seasonal rain in such a way that useful winter rain falls as far north as the Darling Downs in south-eastern Queensland, while eastern New South Wales has an even distribution of rain throughout the year.

In the country dominated by this range system and its water-shed (and particularly on the eastern side of it) are found kinds of mammals, which do not occur elsewhere, such as the true Pademelon Wallabies (*Thylogale*), the Greater Glider and the Koala. However, within this mountain-dominated area (which starts in effect in the Mount Gambier district of south-eastern South Australia) there are still differences between north and south in the species of mammals which live in it because the change from summer to winter rainfall is still there at the extremes and, superimposed upon these environmental factors, there is the historical effect of migration into the area from outside. Thus, some species have moved between this mountain-dominated area and New Guinea at intervals over the last million years, during the periods of low sea-level when these exposed the sea bed across the Torres Strait. Animals in this category include the tree-kangaroos and the cuscuses.

Southern and South-western Australia:

The south-west and the south-east have much in common and the resemblance is particularly marked between the south-west and those parts of South Australia, south-western New South Wales, and Victoria which are clear of the eastern water-shed of the Divide. A large number of species are common to both areas, so the climate over the head of the Bight during the last million years must have been good enough to allow the flow of many (but not all) species across country which they would now find inhospitable. Nevertheless, even during times of climatic amelioration, ecological conditions across the Nullarbor area have not been identical with those at the south-eastern and the south-western ends of it, so that it has acted as a filter, letting through those with less demanding requirements and holding back the more specialized. The result has been that the historic factors which caused the areas to be different before the flow have continued to exert their effects; thus, while the Tammar Wallaby and the Ringtail Possum are found both in the south-east and the south-west, the Honey Possum and the Quokka have never been found in the south-east, yet they are plentiful in the south-west.

One small area in the lower south-west supports a habitat somewhat similar to the mountain-dominated east coast. This is the area of Karri forest which is sufficiently far south to gain the benefit, in summer, of the edge of circumpolar rain systems, which move from west to east to the south of the main continental mass during the summer months.

The Far North:

The Gulf of Carpentaria and the extensive grass plains extending south from the head of the Gulf have tended to separate the northern faunas into two components in much the same way as the Australian Bight separates the southern faunas into south-western and south-eastern elements. The faunas of Arnhem Land and the Kimberley are very like each other but are rather different from that of Cape York, which is dominated by the eastern mountain system and its historic relationship with New Guinea. As in the case of the south-west and the south-east, separation is not complete so that although some species, such as the Little Rock-wallaby and the Rock-haunting Ringtail, are confined to Arnhem Land and the Kimberley, others, such as the Little Northern Native-cat, are found right across the north and extend southwards for a considerable distance, along both the east and west coasts.

The Centre:

The only uniform thing about the Centre is its irregularity both in rainfall and topography; there are ranges, flat grassy plains, salt deserts, stony deserts, densely vegetated sandy deserts, and even deep and relatively humid gorges. These combine to make a very varied ecological situation with a whole host of species, many of which have seemingly rather narrow specializations. In general, though, the fauna of the Centre seems to be made up of a number of discrete groups of species which have different relationships with faunas from outside the Centre.

First, there are species which also occur in the far north. The probable explanation for the occurrence of these species in the Centre as well as the north is that the north contains, among its varied environments, a habitat of rocky country with poor soils in which runoff is rapid and vegetation is sparse. As a result, except when it is actually raining, the animals which live in these places effectively occupy a desert environment which is not very greatly different from that of the Centre—an environment completely lacking in water in the winter but with a violent water supply and high humidity during a short summer period. Species in this category include the Rock-rats (*Zyzomys*) and Red-eared Antechinus (*Antechinus macdonnellensis*).

Secondly, some species from the south-west and the south-east are also found in the Centre, but less is known about these; they seem to be mammals which in their southern homes live in places with well-drained soils. Animals like this, such as the Fat-tailed Dunnart (*Sminthopsis crassicaudata*), the Numbat (*Myrmecobius*), and the

Crescent Nail-tailed Wallaby (*Onychogalea lunata*), do not seem to go very far north into the Centre, and one can only surmise that they may be limited by summer rainfall, which may not suit the southerners even though they can tolerate dry conditions well.

Finally, many of the species of the Centre are found nowhere else—most of these, like the Desert Rat-kangaroo (*Caloprymnus campestris*) and the Mulgara, or Canning's Little Dog (*Dasycercus cristicauda*), are problem species about which we know very little. When the story is fully told of the wonderful biological adaptations which they possess, and which enable them to live under the extreme physiological stresses of the Australian deserts, it will be one of the most remarkable stories of Australia.

The habitat of mammals

From the endpaper map, we can surmise the general patterns of mammal distribution, but this is all. If you are a naturalist looking for a mammal, you must be able to go farther than this and, as well as knowing its distribution, you must, if you are going to be successful in your search, also know its favoured habitat. For example, you might be aware that the distribution of the Swamp Rat (*Rattus lutreolus*) includes Victoria, but there would be no point in your searching for it in the heath of the Little Desert near Nhill in western Victoria, when its habitat is the wet, close-growing vegetation of swamps and gullies.

In Chapters 5–12, where the distributions of the various species of Australian mammals are outlined, you will not only see the familiar patterns comparable with the map, but you will also see a very brief description of habitat in terms of the principal categories of vegetation with which each species is associated. The categories of vegetation used here are mainly those described in *The Australian Environment*, an important and easily obtainable publication. It is an invaluable work which every naturalist should possess.

The principal terms from it, and their meanings, which are used by us here are:

Forest:

Crowns of trees more or less touching or overhanging to give a continuous canopy. Categories of forest are:

Rain forest—a dense forest associated with high rainfall: there is a continuous canopy, and usually several layers of canopy. There are many creepers and epiphytes growing on trees. On the ground there is a thick litter of fallen vegetation and many mosses and liverworts. Decay of the thick litter is rapid because of the very wet conditions. Examples of rain forest are the jungles and scrubs of New Guinea and eastern Queensland, and the antarctic beech forests of western Tasmania.

Wet sclerophyll forest—eucalypts are the usual sclerophyll trees (from the Greek *skleros*—hard, *phyllon*—a leaf): there is a well-developed shrub layer with many of the shrubs being moisture-loving. There is often a second tree layer. Ferns, including tree ferns, are frequently present. Examples of wet sclerophyll forest are the Mountain-Ash forests of Victoria and Karri forests of Western Australia.

Dry sclerophyll forest—eucalypts are the principal trees and there is an understorey of shrubs suited to life under dry conditions. Examples of dry sclerophyll forest are the Box, Stringybark and Messmate forests of Victoria and New South Wales and Jarrah and Wandoo forests of Western Australia.

Woodland:

The dominant trees of woodland are scattered more widely than in forest so that gaps occur between trees or groups of trees; these gaps are at least as wide as the diameter of the crowns of individual trees.

Heaths:

Plant communities with main plants not more than five feet high. Sometimes trees are scattered through a heath which is then described as a tree heath.

Grasslands:

Plant communities which consist principally of grasses although other herbs and perennials are usually present. Hummock grassland of *Triodia* forms the wide spinifex plains of the Centre. In Australia the word 'steppe' is often used to include such grasslands. As shrub steppe, or succulent steppe, it is also used to denote salt-bush plain, blue-bush plain, etc.

In this book you will find the word 'savannah' used as an adjective meaning that grasses are present; for example, savannah woodland, or woodland with a ground flora of grasses.

The effect of man

There is no need to dwell on the effect which man has had on the wildlife of Australia. Before we came, our Aboriginal predecessors left their impression through fire and hunting; they probably caused some species of mammals to become extinct merely because even their relatively minor activities were sufficient to tip the balance of natural selection against the more vulnerable kinds. These less well-adjusted members of the fauna would have included some which, as a result of the long-term climatic changes which followed the last Ice Age, were only just holding their own; others would have been vulnerable to the spread of open country which would certainly have resulted from the use of fire in hunting. Into this category would have fallen many of the animals which the fossil record tells us have become extinct between 12,000 years ago and the time that white men came to Australia. Among them are the great hippopotamus-like *Diprotodon*, the heavily-built browsing kangaroo, *Procoptodon*, and the donkey-sized Giant Wombat, *Phascolonus*.

Our colonial forefathers then came and played their part with more fire, the axe, and flocks. In addition, they created havoc by introducing other mammals such as the cat, the fox, the goat and the rabbit, so that, today, we are left with a fauna which has, almost everywhere, felt the impact of alien elements in the environment as well as the direct effect of man and his actions.

As a result, the environment of almost all Australian mammals has changed. In some cases, the principal change is that additional species have entered it—species that were not there before. Some of these, like the fox, prey upon the native mammals with an efficiency which they cannot combat; and, as a result, some species succumb. Among them, however, are a few which have adaptations which enable them to avoid the new predator—these survive.

But predation is not the only way in which the introductions render the habitat unsuitable for the natives; some will compete with them for food or shelter. Thus it is possible that the absence of the Tasmanian Devil and the Thylacine from the mainland of Australia is due to interaction between them and the Dingo, which was almost certainly introduced by the Aboriginal as his companion. As well as competing with them for food and shelter, and by preying on them, the aliens can, by their mere physical presence, so alter the environment that it becomes unsuitable for native mammals; cattle foul the creeks and break down the scrub along their banks; and the millions of small hooves of sheep break the surface of the soil around shrubs in arid areas, so that the character of the surface becomes changed and the vegetation changes with it.

Add to this the direct effect of man. He has cleared huge areas so that, today, crops and pastures occur where formerly rain forest and sclerophyll forest flourished. Animals dependent upon the sclerophyll forest have gone, and other mammals, better suited to the new environment, have moved in. In a few cases these are native mammals such as the Red Kangaroo, but more often they are aliens.

In looking at the effects of man and his introductions, the naturalist can only recognize that man's use of the land must result in the local extermination of original faunas. Yet their components are not all destroyed because a few species with wide tolerances will survive under the changed conditions; the possum even survives in our suburban gardens. Additional species may even come in because the altered situation suits them better than the old. Eventually our wildlife will settle down in new faunas characteristic of each altered part of the continent. These new faunas will resemble the old native faunas about as much as the present-day English faunas, which inhabit a land of fields, hedgerows and copses, resemble the faunas which the Romans and Anglo-Saxons knew in the same places—when these were uncleared forest. Thus, in the altered areas of Australia, we will have collections of species which are tolerant of man's physical effects such as clearing and burning; among the mammals they will consist of a mixture of a few native species like bandicoots and native-cats, and a number of introductions like the Rabbit, the Hare and the Fox; similarly among the birds there will be the Willie Wagtail and the Magpie, together with the Starling, the Goldfinch and the Blackbird. This may be a depressing thought to some of us, but we must learn to appreciate it: it is happening. Any field guide to Australian mammals written in future will not be complete unless it takes these species, as well as the natives, into account because they are becoming the common components of the fauna of Australia, and one which people will ask to be able to identify at sight in the field.

In this change which is sweeping the countryside of Australia today, is evolution in progress. An evolution similar in kind, but probably magnified in intensity, to the great changes and extinctions of the Upper Pleistocene period when man first began to make himself felt over the wide lands of North America and in the forests of

Europe. The study of these changes today will help us to understand what has happened in the past, and allow the palaeontologist and evolutionary biologist to clothe the dead bones of past ages.

Although these changes are taking place around us, we still have some control over their direction and extent, and one of the aims of the conservationist must be to create sufficient refuges for the native fauna of the continent where the effects of this new evolution shall be minimal. In these reserves representative sections of each major kind of natural Australian community shall be protected, and where necessary managed, in order to retain, in all different parts of this country, examples of what nature herself had created over the millions of years during which the Australian environment has evolved. If we can succeed in doing this, we will ensure that no species of Australian mammal, other animal, or plant, shall ever become totally extinct through man's actions.

Elsewhere in this book (Chapters 13 and 14) we examine what is known of the rarer kinds of mammals in Australia because we hope that through doing it we will stimulate people to find out whether these are in danger. But in focusing attention on the rare ones, let us not forget that even the commonest species today have become exterminated over much of their former distributions—and a few have possibly even become totally extinct.

On evaluating rarity

In trying to evaluate the need for conservative measures, the raw information upon which we operate is, for the most part, observations of animals seen or collected in the wild. If we see an animal very seldom we say that it is rare. If we see it often it is common. Unfortunately, in ordinary, everyday language, when it is said that such-and-such a mammal is rare it is often implied that it is also threatened with extinction. But this is not necessarily true—in fact, the statement can mean quite the opposite—and if we are to propose measures for its conservation we must know why it is rare.

An animal may be rare because we are inefficient observers, or because our paths do not coincide with its paths, or because it is supremely efficient at avoiding us. In general, it is fair to say that rarity must give cause for alarm when it is known that the animal concerned was formerly more common. An animal which is very sparsely distributed over an immense area could be regarded as a rare animal, but each individual of it is only part of an immense population. For example, Mitchell's Hopping Mouse, which occurs thinly between the junction of the Murray and the Darling Rivers in New South Wales and the wheat-belt areas of Western Australia, is seldom seen; but it is probably quite secure as a species. On the other hand, the Quokka, which occurs in small numbers in the valley swamps in the forests of the south-west, was formerly a very common animal which was shot in hundreds. We must be concerned for the future of the Quokka in the south-west.

If rarity in itself is not an indication of vulnerability, how can we assess the vulnerability of a species or even the threat to it of extinction? The answer to this question is probably that vulnerability can only be assessed when we know enough about the habits, distribution, and former population size of the species to enable us to say that it cannot withstand certain kinds of human pressure which it may have to face. For example, a species like the Banded Hare-wallaby (Plate 1) which, today, only occurs on Bernier and Dorre Islands in Shark Bay, Western Australia, is very common there (i.e. it is not rare). But it is vulnerable because the introduction of foxes or stock in large numbers to either of the two islands could put such pressures upon the two populations that the Hare-wallabies could not withstand them and they would disappear as they seem to have done throughout the mainland. This species is vulnerable; but it is not threatened with extinction. Both islands are fauna reserves and are well separated from the nearest fox-inhabited land mass by more than fifteen miles of open water.

From this example we can see that an animal is threatened with extinction when we know or suspect that the pressures on it are exceeding what the animal can stand. For example, a slow breeding species like the Koala was threatened with extinction because it was hunted above the rate at which it could replace itself. It is no longer threatened (see page 12); similarly, a tree-living species would be threatened if agricultural development demanded the total destruction of trees; an obligatorily migratory form would be threatened if fencing reached a density which stopped its free movement.

By inference, a species is almost certainly threatened with extinction if it was formerly common but now, when the same observational methods are used to detect it, it is uncommon.

Later in this book we will discuss what can be done to evaluate rarity and estimate threat and what we, the interested people, the naturalists, and the scientists, can do to combat extinction and perhaps even to reduce the severity of the changes which we are bringing to our countryside. But before we do this let us look at some examples of species which have in a seemingly miraculous fashion been rescued from extinction; let us also see what kinds of native mammals we have in Australia, how they are distributed, and what their habitats are; and then let us try to catalogue the rare ones in order to pinpoint and expose the need where it is greatest.

1   BANDED HARE-WALLABY, *Lagostrophus fasciatus*   Group 6

2

A second chance

ANYONE who has been interested enough to follow the natural history news in the weekly magazines and even the daily newspapers of Australia over the last few years must have noticed repeatedly that such-and-such a species which was believed by everyone to have been extinct has been found alive again. Among those that hit the headlines are the Noisy Scrub Bird and the Dibbler, both of south-western Australia, and the Burramys and Leadbeater's Possum of Victoria.

Have we really been given a second chance to save these species? If so, how has it come about and what can we do about it? Fascinating human stories and romance lie behind each one of these second chances. Here, we can only retell them in broadest outline, but most have been told in naturalist magazines by the discoverers of the animals and all are worth reading and thinking about, not only because they are stories of achievement, but because of the lessons that they teach us. As we read them we will find that some of them really are second opportunities for conserving animals which were nearly lost for ever. On the other hand, more often, these second chances are merely a discovery that an earlier estimate that a species was extinct (or nearly so) was wrong and that we now know otherwise. Yet from each story has come the knowledge that the lost are found and we must now make sure that the second chance, so unexpectedly given, will not be wasted.

THE KOALA   Plate 23

Among the numerous and curious tribes of animals, which the hitherto almost undiscovered regions of New Holland have opened to our view, the creature which we are about to describe stands singularly pre-eminent. Whether we consider the uncouth and remarkable form of its body, which is particularly awkward and unwieldy, or its strange physiognomy and manner of living, we are at a loss to imagine for what particular scale of usefulness or happiness such an animal could by the great Author of Nature possibly be destined . . .

The Koalo* is supposed to live chiefly upon berries and fruits, and like all animals not carnivorous, to be of a quiet and peaceful disposition. Its only enemies must be the Racoon and Dwarf Bear of that country, and from which it can easily escape by climbing; and its appearance at a small distance must resemble a bunch of dry and dead moss. As there are no kind of Tygers or Wolves known as yet, except the Australasian Fox should be reckoned as a Wolf, the smaller animals must be upon the whole more secure than in most other countries.

(George Perry, *Koalo, or New Holland Sloth* in *Arcana; or the Museum of Natural History*, 1811)

All of us, Australians and strangers to this country alike, know the Koala well these days. Its teddy-bear appearance, inoffensive nature, and the numerous toys which can be found on sale in airport lounges and toy shops all over the world have

*The original passage contains Kaolo—an error. Elsewhere Perry employs Koalo consistently.

made it well known—yet few of us know that the story of the Koala is one of the most dramatic tales of a mammal which, not very long ago, was rapidly declining in numbers as a result of man's cupidity and stupidity, and was saved from further decline and possible catastrophe by official action and public horror. It is now secure once more and increasing.

The Koala is a sedentary animal, slow to take evasive action, literally a sitting shot, and, although its pelt was not highly regarded, it was shot and persecuted as a fur-bearer to an extent which seems to have far outstripped its ability to regenerate its population. To give an idea of the magnitude of the kill, in 1908 57,933 Koala pelts passed through the markets of Sydney alone and in 1920–1 205,679 Koalas appeared on the fur market. In 1924, the year in which over 2,000,000 skins were exported from the eastern states, Professor Wood Jones wrote in Adelaide:

> .... many men now living can remember the time when it was by no means uncommon in certain districts of the south-eastern portion of the State. No more than ten years ago Koalas have been killed well within the geographical limits of South Australia. If it inhabits South Australia to-day is rather doubtful, although reliable information would point to the fact that a remnant of the stock may still linger not far from the Victorian border. So far as I know no example of the South Australian race has been examined scientifically, and no specimens seem to have been preserved.

Yet only three years after this, while the news was still fresh that the South Australian population of Koalas had been almost extinguished, 584,738 of them were killed during a single month in Queensland.

The almost incredible account of that Queensland open season of August 1927 has been told in detail by Professor Jock Marshall in *The Great Extermination*. He has told the pathetic story of how the Government of Queensland under Acting Premier Mr W. Forgan Smith, apparently in return for popularity among the electors of a number of country districts—though it is far from certain that the politicians assessed their means of popularity correctly—ignored their professional advisers and allowed the slaughter of a large part of the last undecimated population of Koalas in Australia Then they played it down by crediting £18,897 in royalties to a trust fund for the protection and propagation of native fauna; fortunately, public horror and outcry was such that no State in Australia has ever again allowed the slaughter of Koalas.

It is terrible to kill as inoffensive an animal as the Koala; but the mortally wounded Koala is also so hard to finish off that, in the words of Professor Wood Jones:

> .... horrible cruelties have been committed and recounted by those who have slaughtered them wholesale for the sake of their pelts. Indeed, one may say, on humanitarian grounds, that not only should the slaughter of the Koala for the fur trade be prohibited because the animal is eminently one to protect and not to exterminate, but it should be prohibited because, like the slaying of seals, it is the most brutalizing occupation that a human being can undertake.

And all for the sake of skins which, at the time of the Queensland open season, fetched 56s 9d per dozen.

The story of the recovery of the Koala could not be whiter than the account of the other side is black. In it, individuals like Mr Noel Burnet, the founder of Koala Park in New South Wales, Mr A. K. Minchin of Koala Farm in South Australia, and Mr C. A. M. Reid of Lone Pine Sanctuary in Queensland have shown what private enthusiasm can do if it is not hedged about and prevented by ill-conceived official prohibitions. But most credit for action taken to restore the species as a wild animal must go to the long and sustained effort of the Fisheries and Game* Department of Victoria who between 1922 and the present time have not only succeeded in breeding the Koala in large numbers in the reserves which they stocked on the islands of Western Port Bay, but also in introducing them from these islands into suitable places on the Victorian mainland. In the twenty-five years up to 1957 approximately 7,000 Koalas were liberated in over fifty areas, and in each of six of these (the Mornington Peninsula, Durdidwarrah Reserve, Mount Alexander, the Stony Rises, Mount Cole Forest, and Halls Gap in the Grampians), more than 500 Koalas have been released to form thriving colonies. As Jock Marshall has said, it is heartening to see in some parts of Victoria today the road sign 'Beware of Koalas Crossing Road'.

The success of this work is a stimulating example to conservationists everywhere. Under sound management programmes which have been expanded and developed under Mr A. Dunbavin Butcher, the present Director of Fisheries and Wildlife, and Mr John McNally, now the Director of the National Museum of Victoria, who was Senior Research Officer responsible for the Koala project, the Koala is thriving and increasing. Its future in Victoria is assured. In New South Wales there are a number of thriving colonies particularly in the north-east; and in Queensland, David Fleay believes that it is making a slow recovery from the month of savagery.

The second chance has been taken, and taken well—but the Koala is a large animal, attractive, and easy to become sentimental over. What of the smaller and less obvious ones like some of the mice, and Leadbeater's Possum and Burramys?

LEADBEATER'S POSSUM   Plate 19

. . . the destruction of the scrub and forest in the Bass River has resulted in the complete extermination of one of our most interesting marsupials, the little opossum-like *Gymnobelideus leadbeateri*. There are only four specimens of this extant, and it is extremely unlikely, owing to its very limited area of distribution, that any more will be found.

(Sir Baldwin Spencer, *Victorian Naturalist*, 1921)

The story of Leadbeater's Possum began in 1867 when the great Melbourne Professor of Natural History, Frederick McCoy, wrote about a new kind of small mammal which he had identified in the collection of the National Museum of Victoria. It was a beautiful little animal twelve inches long from nose-tip to tail-tip with a striped face and a long-haired bushy tail. For a new name (and each describer of a new animal must coin one) he combined the name *belideus*, then currently used for Sugar Glider (which means javelin or dart), with *gymnos* (which means naked). In this way, the name conveyed to scientists McCoy's belief that the new possum was related to the Sugar Glider—but was without the flying membranes, which in the Sugar Glider are spread between wrist and ankle.

Through the second name, the professor immortalized John Leadbeater, the taxidermist responsible for preparing the specimens he described.

At the time of his discovery, McCoy knew of only two examples, both from the same place, so in the description he could say no more of it than 'There is only one species known which occurs in the scrub on the banks of the Bass River in Victoria.'

*Now Fisheries & Wildlife.

In this way, and with only this poor ecological information, *Gymnobelideus lead-beateri*, the small possum from Gippsland, was described and named.

In the years that followed, two other specimens of Leadbeater's possum were discovered—but both already stuffed—in private hands. Although these were given to the collections of the National Museum of Victoria, they contributed nothing to our knowledge of the habits of the beautiful little possum and, like the two original specimens, it was believed that they came from the Bass River.

In 1931 attention once more became focused on the species, and hope revived that it still existed, when it was discovered by Mr C. W. Brazenor, the newly-appointed Curator of Mammals of the National Museum of Victoria, that a specimen had been collected many miles from the Bass River at Mount Wills in north-eastern Victoria in 1909. It had lain unrecognized in the collection since then.

As a result of his discovery, Mr Brazenor and his friends searched extensively and fruitlessly for the little animal through the forest-covered slopes between Orbost and Tallangatta in eastern Victoria, and the only consolation they found to counter the gloomy prophecies that it was extinct was the certain knowledge that looking for a small possum in the hundreds of square miles of close mountain vegetation, with no knowledge of its habits to guide them, was like looking for a very small needle in a very large haystack.

The second chance came thirty years later when, in 1961, Mr Eric Wilkinson of the National Museum of Victoria started a survey of the Cumberland Valley, eleven miles east of Marysville and only seventy miles from Melbourne. Wilkinson's survey developed into a project of the Fauna Survey Group of the Field Naturalists' Club of Victoria who have done much good work. One night in April, accompanied by Messrs R. F. Lees and A. Sonnenberg, Eric Wilkinson saw a small grey possum low down on the trunk of a blackwood. In his own words:

It turned and climbed quickly into the upper foliage, hesitated for a while, then jumped across to a neighbouring tree and disappeared from view. When first seen it was thought to be a Sugar Glider (*Petaurus breviceps*) but, as it climbed, a very different kind of tail was seen. Instead of the broad, fluffy tail of *Petaurus*, it had a long, thin tail, bushing out towards the tip, and I suddenly realized that it was probably a Leadbeater's Possum—not in some remote part of the state, but in a well-known tourist spot only 70 miles from Melbourne.

Although the tail was the only diagnostic feature seen clearly, it was enough to suggest a very exciting possibility, one which I had very much in mind while on the way to Marysville some three hours later. A nightjar flew in front of the car and perched in a tree at the side of the road. It flew off as soon as the car was stopped to get a better look at it, but this did not matter in the least, because a pair of eyes reflected in the spot-light beam proved to belong to a small possum very similar to the one seen earlier in the evening. This time a very good view of the animal was obtained, and it was kept under observation for about ten minutes. Although brownish grey in colour, it closely resembled the first one, and the absence of a gliding membrane showed that it was definitely not *Petaurus*. Its size, build and markings readily distinguished it from any other small mammals with which it could possibly have been confused, and it seemed very probable that these two animals were indeed the long-lost Leadbeater's Possum.

Five days later Eric Wilkinson and his friend, David Woodruff, returned to the area and were fortunate enough to see three animals. They succeeded in photographing them and, from that time, there could be no doubt at all of their discovery.

Leadbeater's Possum had indeed been found once more.

Once people knew how and where to look, the possum was found by the Mammal Survey Group, led by Mr W. H. King, in numerous localities extending over some 400 square miles of forest from Lake Mountain, east of Marysville, some thirty miles to Tanjil Bren east-north-east of Noojee, at altitudes between 2,400 and 4,000 ft. It is now considered to be a moderately common species at restricted localities within that range, as it is, for example, on the Upper Thomson River.

Mr R. M. Warneke, the Senior Research Officer of the Fisheries and Wildlife Department of Victoria, believes that the increasing number of locality records since the rediscovery argues that it has built up in numbers and spread out from refuge areas left untouched in the devastating bushfires of January 1939. Further, it has been suggested that the present buoyant state of the population of *Gymnobelideus* has come about because the species is most suited to the environment provided by the present stage of vegetational succession reached by the Mountain Ash forests in their slow recovery from those violent fires.

Since most of the known localities at which the possum occurs are in the catchment areas of three important water storages for Melbourne its habitat is certainly secure for the foreseeable future; we have much to learn, however, before we can be certain whether Leadbeater's Possum depends not only on the existence of Mountain Ash and similar forests, but also upon the existence of suitably sized areas of special stages of tree growth within the forest.

BURRAMYS—A LIVING FOSSIL   Plate 2

The hand of the Lord . . . set me down in the midst of the valley which was full of bones, and caused me to pass by them round about: and, behold, there were very many in the open valley; and, lo, they were very dry.

And he said unto me, 'Son of man, can these bones live?' . . . and as I prophesied there was a noise, and behold a shaking, and the bones came together, bone to his bone. And I beheld, lo, the sinews and the flesh came up upon them, and the skin covered them above . . .

(Ezekiel 37, from verses 1, 2, 3, 7 and 8)

Very few mammals have the strange distinction that they were discovered first as dry fossil bones and only later discovered alive. Such an animal is Burramys, which for seventy years was regarded as a rare Australian fossil. It was even argued over as a missing link between the families of possums and kangaroos until the argument was halted by the discovery that it is a small possum of the high mountains of eastern Victoria.

The strange story of Burramys began on 9 November 1894 when Robert Broom, a young and energetic Scottish doctor, who was in practice in Taralga in New South Wales, went on a picnic to the attractive tourist caves at Wombeyan among the steep mountains not very far from Goulburn.

Broom was a keen naturalist and a very remarkable research worker who was later destined to become the greatest of South African fossil-finders and palaeontologists; he was to find not only Burramys at Wombeyan but later, in the African Karroo, the fossil mammal-like reptiles which are now known to be the links between the true reptiles and the mammals, and eventually, at Sterkfontein, in the Transvaal the skulls and bones of the ancestors of man himself.

2 BURRAMYS, *Burramys parvus*   Group 15

A visitor to Wombeyan today—and it is well worth the journey—will go down a winding mountain road into a narrow steep-walled valley at the end of which the caves lie. Broom explored the caves, and then like many another visitor he climbed up the valley side on to the ridges above the caves; there, in a small depression in the hillside which marked all that is left of an ancient cave-floor, whose walls and roof had long since eroded away, he found some small lumps of a reddish-brown limestone whose surface was speckled with white fragments of fossil bone.

Later, after considerable painstaking work in his laboratory (for the fossils were soft and the rocks hard), Broom obtained sufficient specimens to show that he had discovered a new kind of marsupial about the size of a Sugar Glider but with very strange high, grooved, cheek-teeth of a kind only known in Australia among the Kangaroo family. Broom chose the name *Burramys* (pronounced burra-miss). *Burra* is the Aboriginal name of the district which includes the Wombeyan Caves and *mys* is a common suffix in zoological names which means mouse-like.

In the years that followed, controversy grew over the nature of Burramys. Some scientists had come to believe that it was not a possum at all but a tiny kangaroo; but this was mere speculation because no further specimens had been collected from the fossil deposit and, apart from a few jaws in the Australian Museum in Sydney, the whereabouts of the bulk of the Wombeyan fossils was unknown. Broom had left Australia in 1896 and had taken them with him.

I became involved in the story of Burramys when, as a young research student in Oxford, my advice was sought by Dr Lawrence Wells of the University of Edinburgh who had been a colleague of Broom years before. Dr Wells had come across a collection of fossils in the Anatomical Museum in the Medical Faculty of the University of Edinburgh. In great excitement we found that they were the missing Wombeyan Caves fossils and I was asked by Dr Wells to see if any additional information could be obtained from them. Unlike Broom, who had to rely on chisels and needles to expose the fossils in the rock, I had the advantage of new techniques which had been invented to prepare fossils such as these. Acids freed the bones completely from the surrounding rock, and plastics prevented them from falling apart as the supporting rock was removed; as a result, I was able to satisfy myself that Burramys was really a rather unusual possum related to the Pigmy Possums and to the Gliders and not a miniature kangaroo at all.

There the story would have ended if *Burramys* had remained extinct—as a self-respecting fossil should have—along with *Diprotodon* and *Thylacoleo* and other strange mammals from the ancient past. Imagine my surprise when, thirteen years later, without any warning, I received a telegram from Mr R. M. Warneke which read:

BURRAMYS EXTANT STOP NOT REPEAT NOT EXTINCT STOP LIVE MALE CAPTURED MOUNT HOTHAM STOP AM TRYING FOR FEMALE

Burramys had come to life. The dream dreamed by every palaeontologist had come true. The dry bones of the fossil had come together and were covered with sinews, flesh and skin. Burramys lived. The high-mountain vegetation of the snow slopes around Mount Hotham had hidden it for all these years.

The story of the discovery was a simple accident; one day in August 1966 Dr K. Shortman of the Walter and Eliza Hall Institute and his friend, Mr D. Jamieson, found a small, friendly possum in a ski hut on the slopes of Mount Hotham in the

Australian Alps of eastern Victoria. They decided to bring it back to Melbourne with them and, seeking its identification, informed the Fisheries and Wildlife Department. The Research Officers of the Department had seen nothing like the animal before and asked for an opinion from Mr Norman Wakefield, a well-known Melbourne naturalist who, strange to say, had been working on a Victorian deposit containing Burramys which he had discovered only a few years earlier. From this deposit which contained many of the limb bones of numbers of small native mammals he had been trying, for several years, to isolate the skeleton of Burramys and, through this, to extend my earlier work. His problem was solved.

The Burramys lived in captivity in the laboratories of the Fisheries and Wildlife Department while the officers of the Department searched for a mate for him. They hoped that Burramys could be persuaded to breed in captivity in order to provide the information necessary if proper conservation measures were to be employed. Information on such things as its feeding habits, breeding season, number of young, length of life, and so on were needed. Unfortunately, no sign of any other Burramys has been found and, finally, in May 1967, nine months after his discovery, the only known living Burramys developed paralysis and died.

Burramys is no longer a question mark among the arguments which academics have about marsupial relationships but, in exchanging its status as a fossil for that of a living member of our fauna, it poses an even greater practical question to zoologists as conservationists. There must be other living Burramys somewhere up in the high country. Are they in danger?—Before this can be answered, the population must be found.

THE SCALY-TAILED POSSUM   Plate 3

Long ago, in the early time, Ilangurra had a bushy tail like Burkumba the ordinary opossum. One day when Ilangurra was beginning to climb into a tree, a passing Echidna Koonunginya, in a mischievous mood, seized him by the tail and tried to pull him down. He did not succeed but instead pulled all the hair out of the tail. Thereupon Ilangurra jumped down and in a rage seized Koonunginya and threw him into a prickly bush. Since that day of discord Ilangurra has had a bare and scaly tail and Koonunginya has been covered with spines.

(Legend of the Worora People of the west coast of Kimberley as told to Mr H. H. Finlayson by the Reverend J. R. B. Love)

On a still evening about five years ago I was sitting on the veranda of the house of Mr John Tapper, the Wharfinger of Broome, talking to him about the shells, birds and mammals that interested him so much. At one point, he said casually: 'Can you tell me what kind of possums we might expect to get up here?' He had seen a possum and it was different from the common Brushtail Possum of the south-west which he knew well. I described in turn the possums of Kimberley and the Northern Territory which could possibly be found in Broome and the surrounding Dampier Land; I described the Northern Brush Possum (*Trichosurus arnhemensis*) but he shook his head; the Rock-haunting Ringtail (*Petropseudes dahli*) which had been collected so long ago by the Norwegian collector Knut Dahl in Arnhem Land—the environment could be right but it had never been described away from the Northern Territory—but that drew a blank as well. He said no, they were not like that at all; and he went on to describe for me a possum about the size of a south-western Brush Possum but with short ears, very big eyes, and a long, hairless tail, just like a rasp, which it curled around things.

In these few words John Tapper described the rarest of all the western possums, the almost mythical *Wyulda*, the Scaly-tailed Possum of Central North Kimberley. Until 1917 *Wyulda* was unknown, but in that year the Zoo in Perth received a new possum from Violet Valley Station, near Turkey Creek (between Wyndham and Halls Creek in East Kimberley). This possum lived for a while at the South Perth Zoo and then, after its death, it was examined at the Museum and was found to be a completely new species.

For over twenty-five years nothing more was heard of the species until the Reverend J. R. B. Love, a Presbyterian Missionary well known for his anthropological writings, sent to Mr H. H. Finlayson in Adelaide a second specimen. This had been collected at Mr Love's mission, Kunmunya, 250 miles up the coast from Derby. The local Worora knew it well; they called it Ilangurra and said that it occurred throughout their country.

In the years between Mr Love's discovery and the evening on John Tapper's veranda, the Scaly-tail was only heard of once more. Mr Ken Buller, from the Western Australian Museum, spent June and July 1954 at Wotjulum, a Presbyterian mission to the Worora people approximately half way along the coast between Kunmunya and Derby. While he was there, he collected a female Scaly-tailed Possum which had a pouch-young, but saw no others, and in the following year Mr Athol Douglas, also from the Museum, spent five weeks in the area and searched extensively for the species but without result. It was obviously a very rare animal.

And now I had this species described to me, quite unmistakably, but from Broome, a locality over 250 miles from the closest record and, moreover, right away from Central North Kimberley, over on the other side of Dampier Land. Then John went on to tell me how, one night when he was visiting one of his lights out on the Point, he had found the possum hiding between the gas bottles which supplied the fuel for the lamp. He had taken it home and he and his wife had kept it for a short while. They had then let it run loose into the trees in the garden and had never seen it since.

The knowledge that four specimens of the Scaly-tailed Possum have been discovered many hundreds of miles apart from each other around the edge of one of the most unknown corners of the world, tells us that the rarity of this species may be largely because it lives far from the paths of men—certainly the paths of scientists and collectors. And it is as an example of this sort of rarity that *Wyulda* should be famous to Australian conservationists.

The knowledge that Athol Douglas and Ken Buller searched for it without success, although the Worora of many years ago knew it well and said it occurred throughout their country, possibly indicates that it has suffered a decline but, in December 1965, Harry Butler, who was collecting for the American Museum of Natural History and the Western Australian Museum in North Kimberley, discovered that *Wyulda* was not rare when you know how to look. He found it reasonably plentiful at another locality, Kalumburu up near Cape Londonderry, the northernmost tip of Western Australia. As the result of his efforts he brought to Perth the mother and her pouch-young shown in Plate 3. The young female has now been kept by Ella Fry for over two years. It has thrived on blossoms and fruit and nuts, and has revealed a great deal about the gentle Scaly-tail.

THE DIBBLER   Plate 4

> And no one has a right to say that no water-babies exist, till they have seen no water-babies existing; which is quite a different thing, mind, from not seeing water-babies . . .
>
> (Charles Kingsley, *The Water-Babies*)

Michael Morcombe was writing a book about wildflowers and, because he wanted to illustrate some of the mammals and birds which pollinate some of the spectacular forms, he designed and constructed a special trap which could be fitted over a blossom. With it he hoped to catch a Noolbenger or Honey Possum (*Tarsipes spencerae*—Plate 22), that small narrow-nosed nectar-feeder of the south-western sandplain. The Noolbenger was important for his purpose because, as a result of millions of years of slow evolution, correlated with the prolific flowering of the native flora, it has become one of the most highly specialized nectar-feeding mammals in the world. With its long fingers and toes it holds the slender stems, its long tail grips them, and its long slender face and grooved nose allow it to insert its brushed tongue deep into the blossoms. It is so specialized that its jaws are no more than thin delicate splints of bone and its teeth have almost disappeared.

He set the little trap at various places to the west of Albany along the south coast, but without success; finally he moved it into the waist-high, wind-pruned, thickets immediately beside the little fishing settlement at Cheyne. Beach in the lee of Mount Many Peaks, that lonely little mountain, on which a wisp of cloud continually hangs. He caught no Honey Possums, but instead he got a male and female of a small hairy-tailed marsupial with brown fur flecked with white and with pale rings of fur around its brown eyes. An active and fearless little mammal with a conical snout which has an inquisitive pink tip, the Dibbler had been found again after eighty-four years.

Like the rare Parma Wallaby, the Dibbler is one of the species of marsupial which owes its original discovery to those indefatigable collectors who worked for John Gould. The collector, John Gilbert, discovered the Dibbler in 1838 in the vicinity of the Moore River in Western Australia not very far north of Perth. Nowadays the monastery of New Norcia stands there.

Gilbert got several of the little mammals, both at the Moore River and at King George Sound where Aborigines told him that their name for it was 'Dib-bler'. He learned only a little of its habits; a native at Moore River told him that the Dibbler heaped up a nest of grass and sticks larger than that of the common bandicoot and that this nest was set in a small depression in the ground.

In the years which followed Gilbert's discovery little more was added to this information. A number of collectors working from Albany at King George Sound collected Dibblers and sold them to a number of dealers and museums but, unfortunately, as was often the case with professional collectors of those days, no natural history information, or even precise localities, accompanied them. The most recently caught of these specimens was bought by the National Museum of Victoria in 1884.

Michael Morcombe was not quite sure of the identity of the animals that he had caught, so he kept them for a couple of months, watching them, photographing them, and learning their habits. As a result we now know that they are not strictly nocturnal and that they are out and about as soon as the light becomes less intense towards the end of the day or in dull weather. They love thickets and leaf litter, moving with equal facility through the fallen leaves and twigs, or the twigs and branches above the

4 DIBBLER, *Antechinus apicalis*

ground. They are agile climbers and, when disturbed, drop into the litter and force their way rapidly through it, coming up to look around a little farther on. When provided with a nesting box, they gather leaves and grass and other bits of suitable material and use it to line the chamber. After he had had them in captivity for several months he sent a photograph with a request for information, to Mr Basil Marlow, the Curator of Mammals at the Australian Museum in Sydney, who suggested that I would like to see the animals.

The little animals were undoubtedly Dibblers and, very soon afterwards, Michael Morcombe and I, accompanied by a small party, went down to the original locality from which he had obtained the specimens and, although the season had changed and it was no longer possible to trap on blossoms, we were able to carry out a survey of the small mammals of the area. This survey proved most informative because we were able to show that the area in which the Dibbler lives is incredibly rich in small mammals; moreover, this waist-high coastal sandplain with its stunted Banksias, Christmas Tree, Dryandras and other flamboyant blossoms, has not been examined thoroughly by naturalists in search of mammals for many generations.

At the Dibbler locality, which is within seventy-five yards of the nearest house and standing on a block within a surveyed townsite, we caught another female Dibbler. Her pouch was empty, but she was an adult and before long she had given birth to eight small joeys. She, her joeys, and the first two are now in the capable hands of Dr Patricia Woolley at La Trobe University in Melbourne, where they will be reared in order to establish a captive population which we hope will thrive well enough to provide the breeding information and other knowledge so necessary to provide a foundation for planned conservation.

THE PARMA WALLABY   Plate 5

> This species must not be allowed to become extinct again.
>
> (A New Zealand politician—quoted by K. A. Wodzicki
> and J. Flux, *Australian Journal of Science*, 1967)

The White-throated Parma Wallaby of eastern New South Wales was regarded by most mammalogists, including myself, as extinct. It is a beautiful species with a rich brown back, white throat, and prominent face-stripe and, even at the time of John Gould and his collectors, was not very common. It was found in the rain forest around Lake Illawarra near where Wollongong now stands.

Nearly fifteen years ago I became interested in the Parma Wallaby because I was at that time working on some specimens which Gould had had in his collection; among these was a skull of the Parma. I found that very little was known about the species and, in order to redescribe its characters I obtained the loan of all the known specimens from museums all over the world. There were not many of them—only twelve specimens—and with them no information at all on the natural history of the animals they represented, beyond a note by John Gould that this wallaby loved thickets. At the time that he wrote, he could say nothing more of its status than 'in those extensive brushes it doubtless still exists'.

The Illawarra was an extensive area of rain forest which, since Gould visited it in 1840, has been virtually destroyed and cleared for agricultural purposes. The 'extensive

PARMA WALLABY, *Macropus parma*

brushes' of John Gould are now no longer there. Two specimens were obtained in 1932, however, in the Dorrigo area of north-eastern New South Wales about 300 miles north of the Illawarra district.

One of the results of examining all the known specimens was that I became convinced that the Parma Wallaby was rather like a small Black-striped Wallaby; so, after that, every time I went into a museum I would go to the specimens of the Black-striped Wallaby in the hope of locating further specimens of the Parma which might have with them a little more information. On one such occasion, I found a skin in the Australian Museum which had been sent to Sydney many years ago from New Zealand for identification. New Zealand nowadays is the home of a number of introduced species of wallaby which are regarded as vermin by foresters and land-holders and, in order to gather information on these pests, specimens had been collected and sent to Australia. The skin which puzzled me was without a skull, so I was not able to be certain of an identification; nevertheless it looked rather like a Parma and had been collected on Kawau, a small island off the north-eastern coast of the North Island near Auckland.

For many years Kawau had been the home of Sir George Grey, the Governor of New Zealand who, as an Australian explorer in his younger days, had done much to increase knowledge of the little-known Australian mammals by collecting them and sending them to his zoologist friends in Britain. The little Honey Possum of the south-west was collected by him and its specific name *spencerae* is an honour to Lady Grey, who was a daughter of Admiral Sir Richard Spencer, the Resident at King George Sound from 1833 to 1839. It is not surprising that Sir George Grey, with his interest in Australian mammals, should introduce wallabies into his grounds to keep them in the same way as his countrymen across the world kept deer and hares in their private parks.

I was unable to be certain of the identity of the skin from Kawau so I wrote to my colleagues in New Zealand, Dr K. A. Wodzicki and Dr R. I. Kean, and asked them to watch for the Parma Wallaby on Kawau. Unfortunately, the first trip that was made to see if they were there collected only the Tammar Wallaby, which had also been introduced on to the island. In 1965, however, a further expedition was made by Dr Wodzicki and Dr John Flux and it was found that the Parma really was there. Specimens were sent to Australia to confirm the identification and even modern scientific techniques of blood serum analysis were used to show that the rare species had been found and that they were not merely unusually coloured Tammars.

Unfortunately, the Parma is regarded as a threat to the pine plantations on Kawau and intensive control operations had just been started at the time of the discovery in an attempt to exterminate the wallabies. These have now been halted, but it is far from certain that the population is secure. As Dr Wodzicki has said:

> The irony of the situation is that Australians have been travelling overseas for the past 30 years or so to look at the dozen specimens in museums in Britain, Holland and the United States while on Kawau they have been trying to exterminate the only live ones.

The Wellington correspondent of the New Zealand *Weekly News* estimated that during the past years 3,000 wallabies have been poisoned or shot on Kawau by hunters or business interests who were trying to establish the pine plantation and that probably 2,000 of these animals killed were Parmas.

The rediscovery of this wallaby overseas is perhaps the most remarkable of our second chances; and it is one which has had a most remarkable sequel. Immediately after its rediscovery, a number of Australian zoos and research laboratories imported small numbers of the wallaby in order to study its requirements for conservation and to build up captive stocks for security; many of us hoped that it could be reintroduced into Australia in sufficient numbers to ensure its survival. However, not long after Professor G. B. Sharman established a small number in the University of New South Wales, Mr Eric Worrell told him about a small wallaby that he had caught near Gosford, New South Wales. There seems little doubt that Eric Worrell has now rediscovered the Parma in Australia.

At present we know nothing about the Gosford population. We do not know whether its future is secure or even that the population is large enough to keep itself going. Until we do, we must hope that the New Zealanders can be persuaded to stay their hands over their 'introduced pest'. It may be that a full study of this small wallaby in its 'wild' state in New Zealand will tell us what we can do in Australia to secure the safety of the remaining population. Perhaps, too, the Government of New Zealand might find some method of control which does not involve the total extermination of this historic and fascinating population on Sir George Grey's island of Kawau.

THE GHOST BAT   Plate 51

It would be fruitless, if not impossible, to point out all the peculiarities to be found in the various tribes [of bats] which abound in every country in the world, and differ from each other more in their habits and dispositions than in their exterior form and appearance, which in all of them seem to be equally deformed and disgusting. But we should not from hence conclude, that imperfection and deformity are always in uniform analogy with the notions we have preconceived of what is fair and beautiful. Amidst the infinite productions of creative power, variety of form, difference of faculties, and degrees of utility, are eminently observable; composing one general plan, in which wisdom, order, and fitness, are displayed through all its parts.

(T. Bewick on Bats in *A General History of Quadrupeds*, 1792)

In 1880 a giant white and silver-grey bat, so large that in flight it looks like a seagull, and with long fangs and protruding noseleaf, was described for the first time from south-western Queensland on the Wilson River (a tributary of Cooper Creek). Later, in 1890, 1894 and 1900 additional specimens of it were secured from Alice Springs and from the Pilbara of Western Australia.

The discovery of this giant bat in Australia was of great interest to zoologists because it belonged to the family of so-called false vampires which are carnivorous, and which are known in Africa and Asia. Unlike the true vampires of South America (which feed exclusively on the blood of warm-blooded vertebrates) the false vampires catch and kill many different kinds of animal and some are even fish-eaters. The Australian representative, the Ghost Bat, swoops on and catches, kills, and eats other bats, mice, small marsupials and even lizards and birds.

Twenty-five years after the last specimens were taken, Professor Wood Jones was writing *The Mammals of South Australia* and made an effort to locate the bat in that State, but without success. In writing of this search he said:

The remarkable fact, however, is that at one time it was a common cave species over a very wide area in South Australia. In the Buckalowie and Arkoona Caves in the Carrieton district I found the remains of this bat in large quantities, and from many caves in the Flinders Ranges its remains have been reported, in many cases in the form of almost perfect mummies . . .

In these caves, *Macroderma gigas* appears to have lived for long ages, for the accumulations of guano, which have been exploited commercially, were apparently due to the long occupation by these large bats. During their tenancy they shared the caves with the extinct *Thylacoleo* and *Thylacinus*, whose remains are found in the guano deposits; but even in the Carrieton Caves they long outlived these extinct creatures, for their remains are also found among the most recent deposits, and their perfectly preserved, but fragile, bodies tell of their comparatively recent occupation. *Macroderma gigas* would seem to be an important link in the story of the passing of an archaic fauna from the progressively desiccating interior of the Continent. It lingered on for some time after the heyday of the archaic specialized marsupials was passed. It lingered on until the changed conditions of increasingly desert environment made the obtaining of sufficient food impossible even for a powerful and volant animal . . . For the philosopher, or for the politician, who regards the north of this State as a paradise likely to continue improving, there is this to remember, that within comparatively recent times the increasing desiccation has driven *Macroderma gigas* from Carrieton to Alice Springs—and beyond. It would be a difficult matter to find a place in South Australia to-day where conditions were so luxuriant that sufficient insectivorous bats could be congregated in order to sustain a cave-full of *Macroderma*.

Like many other Australian naturalists today, I learned much of what I know about the fauna of Australia from Professor Wood Jones's *The Mammals of South Australia*. Even though it was published more than forty years ago, we can learn much from it today that no other work can give us, for Wood Jones was a wonderfully observant anatomist. But this section on the Ghost Bat left me with a firm idea which I now know to be wrong; the idea that the Ghost Bat is rare but was formerly more common, and that the evidence indicates that it has a rapidly shrinking range due to increasing aridity.

Eleven years ago, soon after my arrival at the Western Australian Museum, I started to travel, collect, and learn about the mammals of the West. I was introduced to the country and its problems by my colleague, Athol Douglas, who since boyhood had been a goldminer and animal collector among many of its hidden places. Some years before, Athol had become interested in the Ghost Bat when he discovered some in the West Kimberley, and he told me that it was a common form over a considerable area of Western Australia if one knew where to look and how to surprise it at rest. I confess that I greeted his statement with some scepticism—this was before my preconceived ideas on mammal distributions were all upset!

Soon afterwards Athol and I were in the Pilbara, that vast land of spinifex and rocky ranges which extends inland from the coast between Port Hedland and Roebourne. There, in the ranges close to Marble Bar and Nullagine, Athol led me down narrow mine tunnels which had been cut by prospectors into the hills; in them, in the light of our headlamps, we could see the great Ghost Bats slip silently from their roosts at the first hint of danger and fly out through their secret exits into the daylight outside, and then off to an alternative hiding place in a shaft or adit elsewhere. In the tunnels where there was no alternative exit, they even slipped between us and the rocky walls as we picked our way very cautiously past the rotting timbers which held the shattered roof above our heads. We would only know of their passing because they would become entangled in the nets with which Athol normally sealed the entrances behind us. No wonder that they were considered rare; but for these nets, on many occasions we would never have seen the Ghost Bats in the tunnels at all.

The Ghost Bat is now known to occur in many localities north of 28° South, in Queensland, the Northern Territory and Western Australia. Its range has certainly shrunk since the Pleistocene times and this shrinkage is, as Wood Jones pointed out, almost certainly the effect of long-term changes. But the changes may not be those due to increasing desiccation; dry country seems to suit the animal very well. A great deal of study will have to be done before we have the answer to that.

The Ghost Bat is a truly remarkable Australian animal and, despite its somewhat bloodthirsty appearance and rapacious habits, it deserves all the protection we can give it. Because it is so restless at the presence of intruders, it could easily be disturbed so much that its security could be jeopardized. Fortunately, most of the mine shafts which it inhabits today are disused and are very dangerous to enter, so that it is likely that it will continue to exist undisturbed—at least in that environment—providing we do not alter the surrounding country so much that the various species of small mammals, birds and lizards upon which it depends for food become too rare to support populations of Ghost Bats.

3

The names of animals

The biological meaning of the word species in connection with mammals

ONE OF the most difficult aspects of talking about extinction and allied problems in Australia today is that we are still uncertain as to how many species of mammals we have, and where they are found. We are far advanced in knowing the general outlines of our mammal fauna and many species are well understood but, every now and then as we work through our specimens trying to identify some mammal or other, we have a sneaking suspicion that two of the species previously thought to be different may really be one, and that the slight differences which separate them may be no more than differences between individuals. In this category of uncertain species are a number of the so-called species of native mice, and a number of the described species of ringtail possums and rock-wallabies.

Moreover, almost every year, some species not previously known is described for the first time and its name added to the faunal lists—and just as frequently names disappear from the lists as well. Species have not become extinct when this happens, but the names are removed because it is found that they are not species at all.

Throughout the last chapter we made the assumption that the species we were discussing were real entities, and that they have the same fundamental difference between them as exists between cats and dogs, and horses and cows. That such entities (or species) are the meaningful units in the context of extinction is bound up with their ability to reproduce only within themselves but not outside. When a species is down to its last number, it cannot keep itself going by mating with a member of another species. Thus, all horses are capable of mating with other horses, and of producing fertile offspring as a result—but the product of crosses between horses and donkeys cannot reproduce a second generation because mules are not fertile. And cats are not capable of mating with dogs—some species are too different from each other to mate at all.

In everyday conversation, when we talk about such 'kinds' of animals, a biologist would use the word species. It is obvious that much of the interpretation of rarity, and the meaning of the information in this book, must depend upon accurate identification of individual records with species, and that the names given represent real biological species.

When we come to deal with a fauna of wild mammals, as we have done in this book, we find that there is little biological information to help us to decide whether many of the so-called species are real entities. In most cases, the reason we call them separate is because the known individuals seem to be different from those of other species; and it is on these grounds that a newly-discovered species is given a name. Unfortunately because of poor communication between biologists, through incompetence, or for other reasons, it sometimes occurs that a species is given a name on more than one occasion, and the literature contains a lot of such names. Many biologists adopt the procedure that, unless it can be definitely shown that two named species are identical, they should be maintained as two separate species. Other biologists believe that it is more sensible to treat similar specimens as members of the same species unless very strong evidence can be brought to show that their separation into more than one species is warranted on the grounds that the differences between them are constant, and are not merely those which may be sexual characters, the result of slight climatic differences between localities, or even stages of growth or moult.

If in reading this book, you find that numbers of names, well known to you from other works, are missing from its lists, you will recognize that my colleagues and I belong, in general, to the latter group.

If this book had been written only a year or two earlier a number of other species would have been included among the list of rare ones, but they are not here today. In order to illustrate some of the main reasons for the removal of species from the list, the stories of two of these cases are told here. Like many of the stories of the 'Second Chance' they illustrate the theme that changes in the status of named species are often the result of growing knowledge about the animals and are not due to something that has happened to the animals themselves.

To speak of the extinction of many named species is about as sensible as talking about the extinction of the unicorn. It does not exist—and never has done. Such species originate in supposition, remain only in the minds of men, and in their books.

THE KING RIVER PIPISTRELLE

The story of the King River Pipistrelle started for me in 1957 when I first came to the Western Australian Museum. I arrived at a time when most naturalists were gloomy about the survival of many species of mammals. There had not been the money, the equipment, nor the knowledge available for proper searches to be made for mammals for more than forty years and, although enormous strides were being made in our knowledge of the biology of some species like the Quokka, the Euro, and the Brush Possum, very little was known about most mammals beyond the gloomy prophesies in the literature which were mostly the result of the absence of any fresh information for many years.

According to the literature, one of the rarest of all Western Australian Mammals was a little brown bat called the King River Pipistrelle. It had only been collected once and this was by some unknown who had collected and stuffed two small bats; they had subsequently been acquired for the British Museum by Guy Shortridge who had visited and collected mammals in Western Australia in 1905–6 for Mr W. E. Balston.

The two little bats were acquired at King River near Albany and Shortridge sent them to Oldfield Thomas, the great mammalogist who, in common with many of his time, was an indefatigable describer of new species. Thomas recognized immediately that these specimens were something new and gave them the name *Pipistrellus regulus*. No further discoveries had been reported subsequently and, knowing this, I was delighted to discover that the collections of the Western Australian Museum contained a large number of stuffed skins of little brown bats all identified as *Pipistrellus regulus*. Before announcing the rediscovery of this lost species I compared the skins with the description and found that they matched exactly. Unfortunately the description also matched skins of the Chocolate Bat, *Chalinolobus morio*, of which there were no stuffed skins recognized in the collection, but numerous fluid-preserved specimens. It appeared to me that, if a Chocolate Bat were preserved in a solution which kept the ears and lips soft and supple, it would be identified correctly. If, however, it was skinned and dried, then it would be identified as a King River Pipistrelle.

I thought perhaps that I had solved the problem of the rare King River Pipistrelle and sent one of these specimens to the Mammal Section at the British Museum for comparison with the original. I received a reply almost immediately. The skins had been compared and were found to be identical; but the skulls were not.

Without going into details of the skull, the principal difference between that of the King River Pipistrelle and the Chocolate Bat was in profile. The Chocolate Bat has a short little face (rather like a pekinese) and a high bulging forehead; the Pipistrelle had a short face, too, but also a long low head (rather like a ferret); both had the same kind of teeth. So there the matter rested until 1965. In that year I was examining the skulls of some Little Bats (*Eptesicus pumilus*), which are low and long but have less teeth than the King River Pipistrelle, when I discovered one which had the same number and kind of teeth as the King River Pipistrelle. So when I went to the British Museum that year I took it with me and showed it to Mr J. E. Hill, who had made the previous examination for me. I asked him if he would examine the situation again to see if my new specimens could throw any light on the matter.

The results of Mr Hill's investigations have now been published. He has shown, without any doubt, that the two specimens of the King River Pipistrelle consist of two skins of the Chocolate Bat which were formerly stuffed specimens and had been dismounted and made into cabinet skins. One has with it a skull of a Chocolate Bat which consists only of its muzzle and palate. The back of the skull and most of the bulging brain-case which would have shown the characteristic profile of the Chocolate Bat has been cut away in the manner normal in old mounted specimens. The other skin has with it a skull of the Little Bat, *Eptesicus pumilus*, which has never been in a mounted specimen—but this skull has abnormal teeth for a Little Bat, having the same number of teeth as the Chocolate Bat. From that skull, the King River Pipistrelle gained the reputation of having a long, low skull (unlike the Chocolate Bat), while the teeth and skins gave it the dental and external characters of the Chocolate Bat.

The King River Pipistrelle had never existed at all; it was the product of a muddle.

THE WESTERN CHESTNUT NATIVE-MOUSE

On the occasion that John Gilbert collected the Dibbler for John Gould at New Norcia on the Moore River in Western Australia, he also collected some specimens of a small brown mouse which Gould described as new. He called it *Pseudomys nanus*. It has never been seen again in the southern part of Western Australia despite the fact that human habitation is fairly dense and it is not far from a centre of scientific work.

On these data there would seem to be little doubt that the species is very rare in south-western Australia. But is it extinct, or close to extinction? Certainly, it has never been said to have been seen anywhere else, so it would look like it.

A year or so ago Mr Jack Mahoney, of the University of Sydney, and I were working together in the British Museum. He was revising the species of Australian native mice and through the co-operation of our colleagues in museums all over the world, we had not only the collections of the British Museum at our disposal, but we also had very extensive collections from Australia. Some of these we had taken over to Britain for comparison, and others were from famous European collections. And there, with the enormous advantage of being able to lay out all the known material of various species, we found that we were unable to distinguish between *Pseudomys nanus* and the Barrow Island Mouse, *Pseudomys ferculinus*. From there the comparison proceeded; across on the mainland of Kimberley we had the Barrow Island mouse (Thomas had earlier admitted that he could not tell the Barrow Island and mainland Kimberley forms apart) and then into the Northern Territory, and finally, across a gap into Queensland, where there is another, very similar, form called *gracilicaudatus ultra* from Mackay in north coastal Queensland, and *gracilicaudatus* from the Darling Downs to complete the gradual progression. Thus, by doing no more than bringing together from the ends of the earth the described material and re-examining the basis on which the various species had been erected, we came to the conclusion that *nanus* and *ferculinus* were one, and it is likely on morphological grounds that even *gracilicaudatus* belongs in the same species. If this conclusion is correct, and some are sure to say that it is not, the species is not rare or extinct but is a secure and common form on Barrow Island and in the Kimberley, and probably in Queensland as well.

NAMING SPECIES: SCIENTIFIC AND COMMON NAMES

All known species of animals have a scientific name and many mammals have other names as well in the common (or vernacular) languages of the countries in which they occur. Scientific names are international and are always written in Latin or at least in a 'latinized' form using the Roman alphabet. The construction and manner of use of such names is by international agreement between zoologists and is governed by a Code (full title on p. 214). Matters of controversy are dealt with by an elected International Commission for Zoological Nomenclature.

Without going into details here, the Code lays down the procedure which the zoologist uses to decide upon the correct scientific name for any particular species. Since, as we have just seen in the case of *Pseudomys nanus*, a species may actually contain a number of forms thought by earlier workers to be separate, the zoologist, in arriving at the correct scientific name, will often have to select only one of those which were previously used for the separate parts which he is combining under one name.

In formal zoological works it is customary for authors to list all such 'subsidiary' names under the correctly chosen name of each species in what is called 'a synonymy'. I have not followed such a formal procedure in this book, but if you want to find out where many of the species names used by other authors of natural history books are to be found—and hence how the 'species' are grouped together here—you will find many of the more common of them in the index to this book with a cross-reference to the species name which I have selected as the result of applying the procedure of the Code.

The scientific name of a species consists of two parts. The first is the generic name (e.g. *Rattus*). It is always spelled with an initial capital letter and is common to all members of the same genus (i.e. the related group of species). The second part consists of the specific name. It is a single word and distinguishes that species from other species in the same genus (e.g. *Rattus fuscipes, Rattus lutreolus, Rattus tunneyi*). On rare occasions the specific name has the same spelling as the generic name (e.g. *Rattus rattus*) but this has no special meaning. When a generic name is repeated it is often abbreviated only to its initial capital letter (e.g. *R. leucopus*).

In this book I have avoided the use of names for subspecies. These are merely names which are given to geographically separate populations of a single species which, because they have responded physically to particular local conditions, can be told apart from each other. When such populations are given names, a subspecific name for each is added on behind the specific name. Few populations of Australian mammals are well enough known for us to be able to say that they are really consistently different from others in the same species.

The allocation of common or vernacular names to the species of Australian mammals is not as easy as allocating scientific names. First of all there are no agreed rules for guidance beyond common sense and common usage. After all, common names are those used by general people in ordinary conversation and often more than one name is in use for the same species in different parts of the country without any form of ambiguity resulting; and the language is enriched thereby. Among naturalists, however, there are always those who strive for uniformity in common names as well as in scientific names and this book is bound to be criticized for presenting for choice a selection of existing common names for each species in addition to the single unambiguous scientific name.

Generally speaking, common names of mammals can be divided into two sorts—truly common names which arise from the need for ordinary people to communicate about often-seen animals, and 'book' names or 'literature' names which are used only by naturalists who want to use an Anglo-Saxon equivalent for the scientific name of each species. These latter names have only currency among naturalists and are as specialized in their way as the scientific names used by zoologists. Since relatively few mammals are really familiar animals there are few truly common names and even these are mostly used in a generic sense such as Wallaby, Wombat or Kangaroo instead of Red-necked Wallaby or Black-striped Wallaby, and Hairy-nosed Wombat or Bare-nosed Wombat. Some of these truly common names are corruptions of native names, such as the examples given above, but many of them are misapplied names which were first allocated by our zoologically unsophisticated pioneers. Thus, possum is merely opossum (a corrupt version of an American Indian name), bandicoot belongs to a large Indian rat, porcupine to an old-world rodent, and (to widen the field away from badgers, tigers and the like) magpie, robin, willie-wagtail and so on belong to a host of European birds. The early 'misuse' of these names is part of our tradition; only the pedant should be worried or misled by their continued use; and to abandon them on the grounds that our forefathers misapplied them is as ridiculous as to do the same for such words as paddock and creek.

Book names have various origins, but all stem from the common desire to have a name for each species which can be pronounced or remembered more easily by the non-zoologist than the scientific name. Some are coined by the writers of natural-history books without taking any account of local usage and are more often than not near-translations of the Latin scientific names. Examples are Banded Anteater for

Myrmecobius fasciatus, Dog-headed Thylacine for *Thylacinus cynocephalus*, Slender Dasyure for *Dasyurus gracilis*, Gilbert's Potoroo for *Potorous gilberti*, and Bernard's Wallaroo for *Macropus bernardus* (although the last species was in fact named after Mr B. H. Woodward but employed his christian name). Others are descriptively more appropriate than such translations; some of these are newly created in the absence of a truly common name when species are first discovered, but they are more often introduced because some earlier-given common name is regarded as inappropriate after its first use. Examples of this sort are Mountain Pigmy Possum, which some are trying to introduce for *Burramys* which has no truly common name, Fairy Possum which is being introduced in an attempt to replace Leadbeater's Possum, and Echidna which was introduced to replace Porcupine or Spiny Ant-eater. The latter case is a fascinating example of misapplied logic; the introducers of Echidna as a common name took over the old generic name *Echidna*, when this was replaced as a scientific name for the species by the nowadays correctly-used *Tachyglossus*, on the grounds that Porcupine and Spiny Ant-eater are inappropriate for an Australian mammal. Yet Echidna is a Greek word meaning Adder or Viper.

During the last thirty or forty years there has been a conscious attempt to introduce Anglicized versions of local native names as common names. In Western Australia in 1928 a committee of three solemnly sat and produced a list of such names for official 'popular' use; and similar decisions have been made elsewhere. For various reasons, many naturalists have regarded such derived Aboriginal names as being more correct or preferable to other, already existing, common names and quite a number of them have come into very general use as a result. Such names which have become established are Dunnart, Quoll, Wambenger and Numbat. Out of this fashion for native names, however, some extraordinary bastard combinations have arisen; for example 'Northern Quoll' for *Dasyurus hallucatus* (Quoll is *Dasyurus viverrinus*), and 'Red-tailed Wambenger' for *Phascogale calura* (Wambenger is *Phascogale tapoatafa*) and the whole flock of combinations employing Dunnart in the generic sense for *Sminthopsis* (see pp. 122-4).

In short, common names are a sort of game played (sometimes very seriously) by naturalists and, in my opinion, they are best taken and used as they are found without inquiring too closely into their origins or the reasons behind them. They are often colourful and some are truly common. More often than not in such cases local usages are strong and are not uniformly applied from place to place. In this book I have tried to avoid introducing new ones beyond extending the group-use of some of them and I have usually recorded a selection of the existing names—in most cases there are plenty more to choose from. The groups into which the animals are divided approximate, in many cases, to some of the generic usages of common names and it may be that most people will be satisfied merely to use the group names.

We still have a long way to go before we can be certain about the nature of many of our species, and until we have this knowledge names are bound to be a bit unstable. It is sobering to think that we are still only on the boundaries of knowledge in the most ancient of zoological tasks—a task reputedly given by God to Adam.

> And out of the ground the Lord God formed every beast of the field, and every fowl of the air; and brought them unto Adam to see what he would call them: and whatsoever Adam called every living creature, that was the name thereof. (Genesis 2 : 19).

4

The kinds of mammals

The history and composition of the Australian mammal fauna

To a visitor from elsewhere, who is also a naturalist familiar with other faunas, Australia is a strange continent full of unfamiliar plants and animals. It is so strange that many earlier works on natural history describe it as a sort of prehistoric menagerie. Naturalists impressed by tree ferns and lung fish emphasized the view that Australia contained primitive relics and had somehow missed out on gaining the more highly evolved kinds of life.

Today we would regard such a statement as an oversimplification. The species which live in Australia are not relics which stopped evolving; they are modern highly evolved members of long lines of evolution, but many of them belong to groups of animals and plants which once occurred outside Australia. Some have become extinct elsewhere but have continued to evolve to the present day in isolation here.

The Australian mammal fauna is different from those of the other continents because it has been isolated by water since the very beginning of the Age of Mammals, some 120 million years ago. At intervals since that time, a few sorts of mammals have succeeded in getting across the water gaps and establishing themselves in the island continent. The few groups that have established themselves have been very successful; they have evolved in response to the many opportunities and new pressures that the empty continent afforded. As a result, today we have a remarkably unbalanced mammal fauna by comparison with the rest of the world because we have only a few major kinds of mammals, but our mammal fauna is also very strange because those few groups that did get in produced wide radiations of species to fit the many different environmental niches available to them which were unoccupied by other sorts of mammals. Thus, marsupials occupy not only the generalized opossum niche as they do in North America but also the mole niche, the flying squirrel niche, the shrew niche and the wolf niche, to mention but a few of them.

In this manner, very early, possibly even before mammals evolved, the ancestors of our monotremes entered Australia. The modern monotremes, the Echidna and the Platypus, are very unlike each other. As a result we must assume that there was once a monotreme radiation of which they are the only survivors; but our fossil-bearing rocks have so far had nothing to tell us of the paths that the radiation took and what became of their now-extinct relatives. Monotremes occur nowhere else in the world today and their only known relatives are possibly a group of small fossil mammals called Docodonta which lived in Jurassic times about 140 million years ago in Europe and North America.

Also early in mammal history (possibly in late Cretaceous or early Tertiary times, round about eighty million years ago) marsupials entered Australia. Nowadays there is considerable argument between biologists as to where they came from. Some, like myself, suspect that they crossed over from South America (possibly via Antarctica) in the times when Australia was much farther south than it is today and the Antarctic had a warm climate as indicated by the fossils which have been found there; others hold that they island-hopped south-eastwards from Asia along the chain of stepping-stones afforded by the Indonesian island arcs. Today, South America has a rich fauna of marsupials but fossils indicate that it was infinitely richer in the past. In the times when Australia had giant marsupial Diprotodons living their hippopotamus-like lives in the Centre, and the great carnivorous, lion-sized possums, called *Thylacoleo*, haunted the caves of southern and eastern Australia, South America had tiger-sized carnivorous marsupials with sabre-teeth and hosts of smaller species as well, some of which were remarkably like our Australian forms of today.

Early in the Age of Mammals, Europe, too, and North America, had their marsupials. But they were contemporaneous with other mammals and there seems never to have been much of a marsupial radiation in these places; the known fossils are all small carnivorous kinds which must have been rather like modern Australian *Antechinus* in appearance. It is not known whether they ever got into Asia. No fossils are known— but then few fossils are known from southern Asia at any rate, and their absence from there adds little to the argument about the place of origin of the Australian marsupials.

Once here, the marsupials radiated widely in response to the many different habitats available to them. Some, like the native-cat family, the Dasyuridae, remained carnivorous and, until the arrival of the Dingo into Australia with the Aborigines, marsupials were the only predatory mammals. Others, like the bandicoots and the Marsupial-mole, became burrowing specialists, expert at digging up invertebrate prey like spiders and grubs. The most spectacular of the marsupial radiation was the Diprotodonta; this large group provided the continent with all its vegetarian mammals from the small nectar-eating Honey Possum, *Tarsipes*, the arboreal possums and gliding possums, to the antelope-like kangaroos and heavily built hippo-like and cattle-like Diprotodons (which seem to have been made extinct by the Aborigines).

Subsequently, probably not more than some twenty million years ago, and maybe much more recently than that, the first rats and mice came into Australia across the island chains to the north. They settled down as well, and produced the spectacular radiation which contains forms as different from each other as Water-rats, Stick-nest Rats, Tree-rats, Rock-rats, and Hopping Mice. Over the years, the first rat-like invasions were followed by other invasions of rats of the modern cosmopolitan genus *Rattus* which have given us our various species of Bush Rats, some of which (an early wave) belong to peculiarly Australian species and others (recent invaders) which are still members of species to be found today in New Guinea and the Indonesian islands to the north. While these invasions were taking place and the terrestrial species were becoming established, the bats flew in and took up residence, and the seals colonized our shores.

Finally, earliest man came in perhaps as much as one hundred thousand years ago (the oldest dated horizon containing signs which are certainly of human occupation is over twenty thousand years before the present) and, as far as we can tell from the fossils, the Dingo, (possibly being brought in by a later wave of human invaders) joined him something less than ten thousand years ago. Then European man came—but the account of his introductions are not a subject of this book.

How the mammals are described, and how to use the descriptions

The purpose of the chapters which follow this one is to describe the kinds of mammals occurring in Australia. The descriptions of the principal groups are to be found in the pictures—they are so much more expressive than words when you are trying to describe the general features of a mammal; words are mainly used here to tell what we know of natural history and distribution, and to help identify species.

Almost any Australian native mammal is fairly easily recognizable from obvious external characters as one member of some relatively small group of species. Yet to carry the identification farther, so that an individual can definitely be assigned to one or other of the different species of that group, is often very difficult. This is because the most reliable characters used to define and separate species are usually anatomical details of skull or teeth. This book is not written to enable professional zoologists to identify specimens with particular species—other works are designed for this purpose (Appendix II p. 222)—but to introduce the general reader to the kinds of native mammals, to their diversity, to their natural history, and to problems which must be comprehended if the mammals are to be conserved. Accordingly we treat them here at the level at which they may be most readily identified from external characters as whole animals.

To achieve this aim, each group of species is represented by a plate which clearly shows the external form of its members. Some of these groups of species are what biologists call genera; sometimes when the member species of several genera look alike, the group may be broader than a genus. On the other hand where, within a genus, groups of species are clearly distinguishable, that genus may be separated into two or more groups in this book.

Each plate is also accompanied by a section of text which, in addition to brief supplementary descriptive information, contains a summary of, or selections from, what is definitely known of the natural history of its members. This is then followed by a list of the species which we include in the groups, giving appropriate common names, scientific names, distribution and habitat.

In order to make it possible for an animal to be identified with some particular species within a group, each species listed is also described in sufficient detail to enable it to be separated from other species. These external identification characters are given in the list of species immediately following the note on its habitat (where this is known). Often the species of a group are difficult to tell apart by means of such superficial features, but, fortunately, most groups have few species at any one locality so that these descriptions may be found, in practice, to be rather easier to use than would appear at first sight.

Readers familiar with other works on natural history will recognize that many familiar, and much quoted, statements have been omitted. This is partly because we have tried to keep the work brief and some choice has had to be made, but it is also because many of these observations are anecdotal and, when traced back to their sources, are found to be based upon unreliable and unsubstantiated identifications, or even first appear in the literature at secondhand. Lists of publications recommended for further reading are given in Appendix I (p. 213).

The order in which the mammals are presented in this book is not the conventional one in which the primitive kinds are placed first. Here they are given an arrangement which we judge that people will find most useful. We start with the kangaroo family and follow them with possums, the Koala and wombats and these in turn by the bandicoots, then native-cats, rats and mice, bats, the Dingo and seals, and finally the Echidna and the Platypus.

The distributions, and habitats, which are given for the various species may be used as an aid in identification to narrow down the possibilities within a group, because the locality at which an animal is seen is often an important clue to its identity. In using locality data in this way, however, you should always bear in mind that we are still very ignorant about the ranges of a lot of Australian mammals— particularly those of Queensland, the Northern Territory, and north-western Australia —so it is better to be cautious.

If you find a small mammal, and have tried to identify it, do not hesitate to send it to your museum for a second opinion, irrespective of the condition of the specimen. If it is very battered, is smelly, or if it is going to take some time to get there, soak it in methylated spirits or formalin for a few days before sending it; drain it off, enclose it in a plastic bag, and send it to the museum asking for information—or even con- firmation of your identification. As is so obvious from this book, our small mammals are poorly known; by sending in specimens you will add to our knowledge of them. Remember, museum curators do not find it difficult to identify the common sorts, and you will not be wasting their time by doing this. Most curators are pleased to identify many common mammals in the hope of extending knowledge of distribution, or discovering an unusual one. No one will think you ignorant or uninformed because you have found a common mammal; what is common near your home may be very uncommon to a scientist in the museum.

The groups of native mammals and their arrangement

The names of the groups of native mammals which are used in this book, and the way in which they are arranged in chapters, are as follows:

Chapter 5: KANGAROOS AND WALLABIES

Group 1 Great kangaroos
2 Large wallabies
3 Small wallabies and pademelons
4 The Quokka
5 Nail-tailed wallabies
6 Hare-wallabies
7 Rock-wallabies
8 Tree-kangaroos
9 Rat-kangaroos

5

Kangaroos and Wallabies

THE MODERN KANGAROOS and wallabies are a diverse group of ground-living marsupials related to the possums. Of all marsupials they are seen most in daylight and many of the species are quite large. As a result, they are familiar to people and, because some species are thought to compete with domestic stock (and in particular sheep and cattle) for feed, and others are known to damage young trees and crops, they have been more studied than most other Australian mammals.

Today, the kangaroo family consists of three branches. The kangaroos, the various wallabies and the tree-kangaroos make up the principal one. Another is the group of rat-kangaroos which consists in the main of bettongs and potoroos. The third is the little Musk Rat-kangaroo which forms, on its own, a very distinct line. Today it is a small insignificant mammal, but in Pleistocene times (i.e. the period of the last two million years) the Musk Rat-kangaroo had large kangaroo-sized relatives. In addition there was in those times yet another group of heavily-built, browsing kangaroos called Procoptodons which may have gone on all four limbs; they bore the same sort of relationship to the kangaroos as the cattle do to the antelopes and they represented a fourth branch of the family.

Members of the kangaroo family are easy to identify. Most have hind legs that are much larger than their fore limbs and their tails are long and muscular; the tails are used as props when sitting up or moving on all fours and they are important balancers in bi-pedal jumping locomotion. Some of the smaller wallabies and rat-kangaroos move rapidly on all fours, but most members of the family take to their hind limbs when they are in a hurry. Their fore limbs are hand-like and are used in manipulating objects but, as would be expected from their mode of progression, their hind limbs are very specialized with one highly developed main toe (the fourth) which has a strong nail and a long pad underneath the sole extending from the tip of the toe to the heel. On the outer side of the main toe is a smaller fifth toe, while on the inner side are two minute second and third toes which are joined together so that they appear to be a single very small toe with a split nail. These *syndactyl* toes, as they are called, are used as a fur comb. Only the little Musk Rat-kangaroo has a first toe as well.

Like the possums, the kangaroos and wallabies have forward-opening pouches. They have, as a general rule, only a single joey in the pouch at one time. Their teeth are very specialized; the most obvious modification being that there is only a single pair of incisor teeth in the lower jaw which stick forward rather like the blades of a pair of nail scissors. In fact, they look so much like scissors that for a long time a misconception existed among zoologists that they could be used to snip grass in the same manner as scissors; but in modern times cine X-ray studies have dispelled this idea. The three pairs of upper incisors of kangaroos and wallabies are arranged at the front of the mouth like a hoop or vee. Between them there is a leathery pad against which the procumbent lower pair rest or can be pressed when pulling vegetation. Like other grass-eating mammals, there is a wide gap between the hindmost upper incisor and the molars (which grind up the food); this gap is probably used to allow the tongue to manipulate food such as leaves and grass.

This characteristic arrangement of teeth in which one pair of procumbent lower incisors opposes not more than three pairs of upper incisors is called *diprotodonty*. It is possessed by the kangaroo family and by possums, the Koala, and by wombats. In combination with *syndactyly* (see above), it provides certain recognition of the families which together make up the great group of Australian marsupials called the Diprotodonta.

Group 1   Great kangaroos   Plate 6

The kangaroo, to most people, and certainly to most Australians, is the typical mammal of Australia. Indeed, to many, the word *marsupials* means *kangaroos*. However, despite the fact that they are not the only mammals with pouches, they are still very remarkable. The biggest of all living marsupials, kangaroos are the marsupial response in evolutionary adaptation to the open woodland and grassland habitat; an environment which, in placental mammals, has resulted in the evolution in other continents of horses, deer and antelopes. Moreover, kangaroos, like their evolutionary parallels, are fleet of foot and long in limb although, in their peculiar way, they move only on their hind limbs when they are in a hurry.

Until recent years the kangaroo has scarcely been of economic importance except in places where numbers were high enough for them to be a pest. Yet their equivalents in other continents are of vital importance to Man; they have been domesticated by him, and have been brought by him to Australia. In the early days, settlers depended upon kangaroos for flesh but, with increasing production of beef and mutton, kangaroo meat became rarely used for human consumption and only skins retained some importance. In recent years, however, the pet-food trade has resulted in an enormous increase in kangaroo shooting, and one of the greatest problems of wildlife management in Australia today is to know whether the present rate of killing is greater than the populations of kangaroos can stand.

Figures for the crop of great kangaroos from 1954 to 1965 for Queensland show that the harvest for the period 1961–5 was an increase of some sixty per cent on that of the previous five years, and the total crop now exceeds one million per year. The production figures given by Roff and Kirkpatrick of the Queensland Department of Primary Industry of the kangaroo harvest in Queensland from 1954 are:

1954	199,731	1960	748,496
1955	260,727	1961	457,802
1956	316,339	1962	541,720
1957	589,159	1963	626,618
1958	273,897	1964	1,106,542
1959	963,667	1965	1,154,543

In the absence of knowledge of the permissible crop which will allow a sustained yield, the actual figures of marketing are alarming. This fear is supported by the work

6   WESTERN GREY, *Macropus fuliginosus*

of Dr H. J. Frith and his Division of Wildlife Research, C.S.I.R.O., who have shown that in one area of 1,000 square miles of inland New South Wales during the period from April 1960 to January 1963 the population density of kangaroos has fallen sharply by seventy-five per cent, and in another area of 4,000 square miles between 1964 and 1966 it fell by sixty-five per cent: there seems little doubt that these reductions are due to the increase in killing. Research workers have estimated that in Central Australia, in the year 1961 alone, seventy per cent of the population was shot.

There are five species of Great Kangaroo: two of these, the Eastern and Western Grey Kangaroos, are so alike that they were only separated with certainty as a result of studies on breeding biology and blood characteristics in 1966. All species have very marked environmental preferences, thus Grey Kangaroos dwell in eucalypt woodland and sclerophyll forest; the Wallaroo or Euro haunts rocky places and, in arid parts of the inland, is characteristic of spinifex-covered tableland country, breakaways, and rocky hills; the Red is generally found where there are grasslands; and the Antelope Kangaroo inhabits northern savannahs and grasslands.

A great deal is now known about many aspects of the biology of the Greys, the Euro and the Red as the result of research stimulated by their suspected role as competitors with stock in the pastoral industry and, more recently, as the result of fears that populations are declining.

Of the great kangaroos most is known of the Red Kangaroo from the work of Dr H. J. Frith and his colleagues. This species is the kangaroo of the inland plains and occurs where the annual rainfall is fifteen inches or less. It is primarily a kangaroo of open country, but its small parties may be found in all habitats except the most dense scrub. Males may weigh up to 180 pounds but females seldom weigh more than sixty pounds. Males become fertile at about twenty-eight months and remain so through their lives but there may be periods when they cease breeding because of particularly adverse conditions. Some females studied reached sexual maturity before they were eighteen months old, but most did not do so until they were about thirty months old. Environmental conditions appear to play a considerable part in determining the age at which Red Kangaroos reach maturity. Like the males, females also cease to breed under particularly adverse conditions.

Pregnancy lasts for thirty-three days and within a few days of giving birth all females mate again and the second embryo which results from this mating remains dormant in the uterus as a kind of 'reserve'. More than sixty per cent of females examined in the field by Dr Frith's team have been found to be carrying such reserve embryos. Delay in the development of the reserve embryo is caused by the presence of the first joey in the pouch. It is born within a day of the first joey vacating the pouch (unless it dies this occurs at about 235 days after birth). The joey first starts to leave the pouch for short periods at about 190 days of age. Even after the first joey leaves the pouch for the last time, it continues to suckle from outside until it is about a year old. During this period the female produces milk for both young and, in addition, she is by then carrying a further reserve in her uterus.

The mortality of joeys in the pouch varies under different conditions; when these are harsh about sixteen per cent of large pouch joeys die. Similarly conditions affect young which are leaving the pouch; when these are favourable about fifteen per cent of joeys die, in a mild drought about eighty-three per cent die, but in a severe drought none survive and, moreover, no reserve embryos are carried.

As a result of this pattern of reproductive behaviour the Red Kangaroo is marvellously adapted to the fluctuating conditions of our inland environment. In all but the most severe droughts there is a succession of young coming into the pouch; under poor conditions, when females are unable to provide adequate milk, pouch-young die and are immediately replaced; under better conditions joeys complete pouch-life and, if they leave it at a time when they can find sufficient food they survive—if they do not they perish; if conditions are too bad, all young are lost and no more are produced—but with the advent of the first green shoots all females come into breeding condition again. In ideal seasons a female can produce four young in three years.

Research has shown that the widespread belief that sheep and kangaroos compete for pasture is not well founded and needs much more accurate investigation. For example kangaroos, Reds and Greys, in south-western Queensland have been shown by Griffiths and Barker to be highly selective grazers, Greys primarily feeding on grasses, and Reds taking a percentage of dicotyledonous plants but neither preferring the same plants as sheep. Weight for weight, kangaroos do not eat more than sheep and in good years scarcely compete with them. They never eat mulga which is the most important sheep crop in south-western Queensland. In bad years it is probable that there is competition between sheep and kangaroos because shortages may drive both out of their preferred grasses and other plants and cause them to compete with each other for what ground feed is available.

Research has also revealed that by man's introduction of sheep into Australia, and his management of them by allowing them to run in largely unfenced, open pastures, the Red and the Euro may have been favoured by him through making the environment even more suitable for kangaroos than it was previously. In this process woodlands and salt-bush or blue-bush plains have been altered to grasslands while in spinifex country (where more palatable grass has decreased due to grazing pressure from sheep) the spinifex-feeding Euro has been able to expand its populations as the result of the provision of watering points.

At present the kangaroo industry is an extremely valuable one. Frith and Calaby in *Kangaroos* quote the following annual values given by the Customs Department (which do not take into account any local sales for pet meat etc.):

1960–1	A$1,619,968	1963–4	A$2,267,680
1961–2	$1,082,182	1964–5	$3,046,738
1962–3	$1,092,384	1965–6	$2,681,130

The population of kangaroos is a resource which is being exploited without any official policy to conserve it so that it will give a sustained yield. In places where kangaroos really do compete with stock their numbers must be reduced by overcropping, but in other places the crop must be controlled or the resource will be brought so low that it will be lost and these wonderful animals be reduced, like the American Bison, to a small remnant confined to reserves.

Weight for weight the Red Kangaroo and the sheep eat approximately the same quantity of feed, yet kangaroos convert grass into meat more efficiently. Moreover, they do not directly compete. A kangaroo is fifty-two per cent meat; a sheep twenty-seven per cent. For these reasons alone, if for no other, it is surprising that Australian pastoralists have not agitated for cropping rights over the kangaroos which live in their pastures; and also that we do not already farm kangaroos in mixed communities with domestic stock—or at least control their harvesting as a valuable resource. Under certain conditions kangaroos may even be worth more than sheep. Dr H. J. Frith has pointed out that during the recent Queensland drought sheep were worth 30–40c apiece while a large kangaroo would fetch $2.50.

The Species Group 1

GREAT GREY, FORESTER, *Macropus giganteus.* Tas., Vic., eastern and central Qld, N.S.W. (where there is an area of overlap in south-west N.S.W. with the Western Grey); forests or woodland.
Recognition: silvery-grey coat; travels with fore quarters low, head high, tail curved upwards and going like pumphandle; females and males same colour.

WESTERN GREY, BLACK-FACED KANGAROO, MALLEE KANGAROO, SOOTY KANGAROO, *Macropus fuliginosus* (Plate 6). South-western N.S.W. and contiguous Vic., S.A., Kangaroo Island, W.A., forests or woodland.
Recognition: light grey-brown to chocolate, never with silvery sheen; travels with fore quarters low, head high, tail curved upwards and going like pumphandle; males with strong smell; females and males same colour.

WALLAROO, EURO, HILL KANGAROO, BIGGADA, *Macropus robustus.* All over Australia (except Tas.) where there are gullies, rocky outcrops and ranges.
Recognition: hair long and coarse; hands and feet dark, tail darker at tip, general colour from red-brown to very dark blue-grey (looks black) and to fawn; thick-set, travels stiffly upright with hands close to body; females markedly paler than males of same district.

RED KANGAROO, PLAINS KANGAROO, MARLOO (female sometimes called Blue Doe or Blue Flier), *Megaleia rufa.* Australia, mid-latitudes wherever there are extensive grassy plains.
Recognition: hair velvety; black mark through white mark at sides of muzzle (or black mark through white splash), tail paler at tip; general colour from red to grey, females white below; travels head down; males and females usually different colours in same district

ANTELOPE KANGAROO, *Macropus antilopinus.* Northern Qld, N.T., and North Kimberley of W.A.; plains and broad valleys with open savannah woodlands.
Recognition: hair fairly long but thin, hair around digits black, none or almost no black on tail-tip; males reddish-tan, paler below, females the same or bluish-grey, or tan with bluish-grey head; most likely to be confused with *M. robustus*, but does not travel stiffly upright.

Group 2 Large wallabies Plate 7

The large wallabies, as a group of species, are little different from the great kangaroos except in size. Like their even smaller relatives (Group 3), they are merely a group of kangaroo-like species which, by virtue of their environmental specializations, are somewhat smaller in size. The large wallabies find shelter in dense forests, or in woodland, or in the scrub along rivers in otherwise open country. One of them, Woodward's Wallaroo, is highly specialized in the same manner as the larger Wallaroo, or Euro, and lives in rocky places.

In some parts of Australia, and in particular where great kangaroos are rare, such as in Tasmania, or in the Kimberley, the local people may refer to large wallabies

WESTERN BRUSH WALLABY, *Macropus irma*

Group 2

as kangaroos. Like the great kangaroos, the large wallabies carry only one young in the pouch at a time; they are also grass and leaf-eaters. Because of their dependence upon thickets for cover, they appear to be very vulnerable to the effects of clearing and to the introduction of stock which break up this type of cover. However, providing that there is sufficient cover left for shelter, patchy clearing for cattle raising, as has occurred in the area of the upper Richmond and Clarence Rivers in north-eastern New South Wales, actually favours some of the species. This ability to utilize cover from which to come out and graze and feed in adjacent lands is a feature of kangaroos and wallabies that is well known to farmers. Moreover it makes populations difficult to control when they are causing damage. To the conservationist, it means that expensive methods of confining them to reserves must be employed unless adjacent farmers are to suffer. The Black-tailed Wallaby in Victoria, and the two wallabies in Tasmania, are examples of this group which present a serious problem to conservationists of trees and animals alike.

In some places, under good climatic conditions, populations of wallabies may be extremely dense. In Canterbury, New Zealand, where two does and a buck of the Red-necked Wallaby (*Macropus rufogriseus*) were introduced in 1870, control by Government departments between 1948 and 1965 resulted in over 70,000 animals being shot, not counting the many thousands shot by private hunters during the same period; and the control shooting still goes on. It is not known what numbers are killed by poison in the regenerating *Eucalyptus* forests of Tasmania; there they are a serious problem in regenerating stands of *Eucalyptus regnans* until the trees reach a height beyond which the wallabies cannot do damage. In sugar-cane districts of Queensland a bounty system is in operation for Sandy and Whiptail Wallabies. In 1956–60 61,287 bounties were paid, and in 1961–5 69,027. Numbers of the same species taken for the skin trade in other parts of Queensland in the same years were 68,737 and 43,670.

All the large wallabies except Woodward's Wallaroo and the Swamp Wallaby of eastern Australia are obviously closely related and are put by zoologists into the genus *Macropus* (meaning big feet). In this genus are also some of the small wallabies and all the great kangaroos but the Red. Little is known about the relationships of Woodward's Wallaroo, or the Swamp Wallaby. The latter is usually placed in a separate genus *Wallabia*.

The Species   Group 2

Most of the large wallabies have well-marked cheek-stripes and hip-stripes; these, together with a stripe along the back when present, are useful recognition features.

RIVER WALLABY, JUNGLE KANGAROO, SANDY WALLABY, AGILE WALLABY, *Macropus agilis*. Kimberley, N.T., Qld; river country—shelters in woodland, or gallery forest, and feeds in savannah or grassy plains.
Recognition: Uniform golden or sandy brown, paler underside, distinct white hip-stripe and cheek-stripe; tail thin; holds head high.

RED WALLABY, EASTERN BRUSH WALLABY, SCRUB WALLABY, RED-NECKED WALLABY, BENNETT'S WALLABY, BRUSH KANGAROO (Western Vic.), KANGAROO (Tas.), *Macropus rufogriseus*. Eastern Qld, eastern N.S.W., Vic., south-eastern S.A. and Tas.; inhabiting woodland, forest edges, and coastal scrub.

Recognition: face-stripe whitish, indistinct; shoulders and rump reddish (shoulder patch particularly noticeable); general colour greyish or reddish-grey; end of tail and digits dark.

BLACK-STRIPED WALLABY, SCRUB WALLABY, *Macropus dorsalis.* Eastern N.S.W., and eastern Qld as far north as Rockhampton; rain forest, lantana thickets, brigalow thickets, and shrubby thickets in gullies.
Recognition: distinct dorsal stripe down neck along centre of back, distinct curved white hip-stripe; shortish tail which is sparsely haired and scaly.

WHIPTAIL, PRETTY-FACE WALLABY, GREY-FACE WALLABY, BLUE-FLIER, GREY-FLIER, FLIER, *Macropus parryi.* North-eastern N.S.W., eastern Qld; *Eucalyptus* woodland with grasses, usually on slopes and hill tops.
Recognition: tail very long and slender, very marked white face-stripe, colour pale grey with very noticeable white mark starting on hips and going round under tail; upright when hopping with tail curved upwards.

WESTERN BRUSH WALLABY, *Macropus irma* (Plate 7). South-western Australia; dry sclerophyll forests, shrub and tree-heath, particularly on forest edges and 'sandplain'. It occurs in sandplain as far east as Esperance.
Recognition: distinct white face-stripe and black and white ears, hands and feet black, faint barring sometimes visible on back, tail crested with black hair; head held low when hopping.

TOOLACHE (toe-lait-shée), *Macropus greyi.* South-eastern S.A. and Vic. contiguous with it (generally regarded extinct throughout its range).
Recognition: as for *M. irma*, but tail crested with pale hair.

WOODWARD'S WALLAROO, BERNARD'S WALLAROO, *Macropus bernardus.* Interior of Arnhem Land; rocky country.
Recognition: among rocks; males a dark sooty-brown (looks black), females paler.

SWAMP WALLABY, BLACK WALLABY, BLACK-TAILED WALLABY, *Wallabia bicolor.* Eastern Qld, eastern N.S.W., Vic., south-eastern S.A. Dense moist thickets, particularly in gullies. Open forest with patches of dense cover.
Recognition: in northern parts of range back and head reddish grey-brown with orange belly; in southern parts of range brownish or greyish-black all over with greyish sides and sometimes with brownish-red belly or underparts; face-stripe and shoulder-stripes generally rather indistinct; hops heavily with body well bent over and head held low; because of long dark hair on tail it does not appear to taper like that of other wallabies.

Group 3 Small wallabies and pademelons

Plates 5, 8

The small wallabies all look rather like miniature kangaroos but their homogeneity ends with outside appearances. They actually belong to two separate genera. First of all there are the two species which are merely small members of *Macropus* like the grey kangaroos, the wallaroos, and most of the large wallabies. They are the miniature members of that genus which are specialized for dwelling in thick cover. The other genus of the group is called *Thylogale* and is related to the rock-wallabies. These also dwell in thick scrub, especially in rain forest.

The two *Macropus* wallabies are the Parma Wallaby (see p. 22) and the Tammar. As far as is known the Parma is entirely restricted to rain forest in New South Wales and is extremely rare. The Tammar differs from the other small wallabies because it is specialized for life under conditions of much less rainfall than would support the rain forest or even wet sclerophyll where the others are found. Mr J. E. Kinnear and his colleagues have recently shown that Tammars are able to survive on food containing little water; moreover they are able to obtain water from the salty juices of succulent plants and even by drinking sea water. It is possible that the ability of the Tammar to survive on the hot and arid islands off the Western Australian and South Australian coasts results from this capacity. Counting the populations of the mainland of South Australia and of Western Australia as isolated from each other, there are eleven separate populations. Broken up distributions of this sort are ideal material for evolutionary studies because each isolate, although genetically similar to the others, has to contend with rather different environmental conditions and is likely to exhibit differences in structure, physiology or even behaviour; but, as yet, no comparative studies of the biology of the populations of this species have been made. Two of the populations are thought to exist no longer.

In the South West of Western Australia the Tammar finds shelter in thickets formed from the early regenerating stages of Tammar Scrub (*Casuarina*) until it becomes too tall. At present little is known about other components of its shelter but the conservation of a mammal species which requires a particular stage of forest growth to maintain it poses a difficult problem in environmental management. A similar problem may be posed by Leadbeater's Possum (see p. 14). Unfortunately, the Tammar occupies country in the wheat-belt and, apart from land in forestry reserves and a few small fauna reserves, little uncommitted virgin land remains. At present problems of its management are being studied at the Tuttanning Reserve near Pingelly in Western Australia.

The *Thylogale* wallabies or pademelons are poorly known and, except for the Tasmanian Pademelon or Red-bellied Wallaby, have been little studied. The Tasmanian species is better known because it poses a problem in forest conservation; moreover, it occurs in very large numbers in some farming areas. It is economically important to the fur trade; although in pest control operations many animals are merely killed and not skinned. Between 1923 and 1960 two and a half million skins were taken in Tasmania.

The Species   Group 3

In the small wallabies the hind foot, from the base of the nail of the central toe to the heel, does not exceed six inches in length.

TAMMAR, SCRUB WALLABY (S.A.), *Macropus eugenii* (Plate 8). South-western Australia (including Houtman Abrolhos, Recherche Archipelago), S.A. (mainland, some offshore islands including Kangaroo Island); dense thickets in dry sclerophyll forest.
Recognition: no distinct face-stripe, faint dorsal stripe, shoulders often reddish; general colour greyish-brown.

PARMA WALLABY, WHITE-FRONTED WALLABY, *Macropus parma* (Plate 5, p. 23). Eastern N.S.W.; rain forest, scrubs.
Recognition: faint dorsal stripe descending no farther than mid-back; upper lip white; throat, chest and belly white; general colour dark brown.

TASMANIAN PADEMELON, RED-BELLIED WALLABY, *Thylogale billardierii*. South-eastern S.A., Vic., Tas., islands of Bass Strait; low scrub.

8 TAMMAR, *Macropus eugenii*   Group 3

Recognition: smallish rounded ears hidden in fur; no markings; dark brown, little contrast back and belly.

RED-NECKED PADEMELON, *Thylogale thetis.* Eastern Qld, eastern N.S.W.; rain forest and wet sclerophyll forest.
Recognition: reddish shoulders and nape of neck (latter very obvious); general colour brownish-grey, no hip-stripe.

RED-LEGGED PADEMELON. *Thylogale stigmatica.* Eastern Qld, eastern N.S.W.; rain forest and wet sclerophyll forest.
Recognition: reddish marks above heel and about face; yellow hip-stripe (often faint); general colour brown tending to reddish.

Group 4   The Quokka   Plate 9

The Quokka of the South West of Western Australia is one of the smallest of wallabies. Only a little larger than a domestic cat, it is a familiar animal to visitors to Rottnest Island near Fremantle, where it can be seen any time in daylight around the holiday settlement. To the conservationist it is a very important animal because it was the first Australian marsupial to have its biology studied intensively; and, as the result of the pioneer studies of Professor H. Waring, Professor A. R. Main, and their associates, at the Department of Zoology of the University of Western Australia, our knowledge of it has become the foundation of almost all modern work on Australian marsupials. Many of the senior biologists working on marsupials today can look back on what the Quokka taught them when they were young research workers.

One of the earliest discoveries made as a result of work on the Quokka was the process of delayed embryological development in marsupials which has been described already in this book in connection with the Red Kangaroo (p. 42). Another was that this marsupial herbivore employs (as do the non-marsupial ruminants: sheep, cattle, etc.) colonies of bacteria in its gut to digest the cellulose of the otherwise indigestible plant cell-walls; the bacteria break down the cellulose and the animal absorbs the by-products of this; it digests the bacteria as well. But this is not all that makes its digestive processes remarkable for, under appropriate conditions, the Quokka can utilize some of the urea, which is a waste by-product of its own energy processes, to alleviate protein deficiencies in the diet; Professor Main and his students have shown that the Euro and the Tammar also do these things. The knowledge that several species behave in this way helps us to recognize how very efficient the marsupials of the kangaroo family are in utilizing the poorly nutritious vegetation upon which many of them feed; and, as we have already seen in the case of the Red Kangaroo (p. 43) they are well suited for life under the unreliable, and often harsh, seasonal conditions through which they live.

Until the mid-1930s the Quokka was a very common animal in the South West where it occurred in swampy thickets; Quokka-shooting was even a familiar sport. The species also occurred on Bald Island, near Mount Many Peaks, and on Rottnest Island, near Fremantle, where the Quokka was first seen by the Dutch navigator Volckersen as early as 1685. Today it is rare on the mainland where it is only known

9   QUOKKA, *Setonix brachyurus*   Group 4

with certainty in a few swampy valleys in the Darling Range close to Perth; but it is still numerous on the two islands.

Most biological studies have been carried out on the Rottnest population which lives under rather extreme conditions of aridity for an animal which apparently loves swampy thickets. There Quokkas eat a variety of coastal sand dune vegetation, and wattle is both a favourite food and an important cover plant. Quokkas can climb to reach twigs growing on branches up to five feet from the ground.

During the wet winter months, and in mild summers, there is ample water and food for the Quokkas; but in harsh summers the population can get little water and approaches a condition close to starvation as a result of the low quality of the diet available. During these months of near-starvation, the Quokka is able to maintain its water-level through behavioural and physiological adaptations similar to those found in desert animals. Such adaptations are rather unexpected in an animal which naturally, in its mainland swamps, lives in what is, to superficial appearances at any rate, one of our least desert-like environments. On Rottnest Island, where there is virtually no free surface water at all in summer, parts of the Quokka population can reach a few soaks or fresh-water seepages along the edges of the salt lakes. Animals have been recorded as moving up to 2,000 yards to the seepages and it has been demonstrated that they can even find their way home over distances of over 3,000 yards; yet Quokkas appear normally to have rather restricted home ranges. Quokka tracks radiate outwards in all directions from around the soaks and seepages and the populations are very dense there in the summer months from about the end of November to the end of March when the rains come to bring fresh herbage and water. During the dry months Quokkas living too far away to reach the soaks do without water.

Unless they are in captivity, or are able to supplement their diets around the island settlement or the rubbish tips, females cease to be in breeding condition in late winter and commence again in summer; and the first pouch-young of the new season appear in autumn. They are capable of breeding at about eighteen months of age but, in the wild on Rottnest, it is unlikely that they carry their first young until their second autumn which means that they could be up to twenty-three months old at that time.

Pregnancy in Quokkas lasts for twenty-seven days and they next mate one day after giving birth; the second embryo is held in the uterus in delayed development as long as the first joey is suckling in the pouch. If this joey dies, development of the second embryo continues, but, if the one suckling in the pouch lives, the second in the womb degenerates and is lost when the female enters her summer non-breeding spell in October. In the Tammar a delayed embryo may be held dormant for eleven months from the end of one breeding season into the next.

Apart from the population at the western end of Rottnest Island (a population on a peninsula which is separated in practice from those of the rest of the island by the isthmus of Narrow Neck) little is yet known about the actual size and density of the populations on Rottnest. Like the populations of many islands, it is certain that they fluctuate widely and Professor Main and his students have established that on West End the normal population of around 600 individuals may be reduced in a harsh season to about one-third and subsequently regain its numbers. Restoration of the population occurs despite the rather high death-rate which seems, from the studies of Dr G. M. Dunnet, to be usual. He has shown that, in the part of the population he studied over the years 1956–8, out of 1,000 pouch-young only 238 females would survive to their first year of breeding; therefore these must have an average life of five years to maintain the population alone. For the population to be able to increase from a low level must indicate that the death rate is reduced at such a time.

Rottnest is an island playground for the people of the metropolitan areas of Perth and Fremantle and, as a result, it is being developed for recreation. In planning for the future the requirements of the Quokka, and the plants on which it depends, must be known if it is to continue to exist on the island; on present information, it will require numbers of areas, at least of the size of West End, to be left untouched and undeveloped. But, even this alone will not be enough. The area of West End is sufficient for its population of about 600 animals because it contains at present an adequate proportion of a variety of habitats, and of mixed food and cover. If other areas of the island are to be selected as equally suitable, we will have to learn what these requirements are: and how they can be kept in balance despite the destructive actions of the Quokkas themselves and the natural changes which occur in the environment as plants get older and as the populations of animals and plants are affected by climatic fluctuations.

The naturalists of Australia owe much to the little Quokka of Rottnest. It would be a tragedy if it were to be squeezed out of existence simply because we had not discovered until too late how much we could reduce its environment with safety; and also what we have to do to manage the reduced environment and its populations in order to compensate for the consequences of our actions.

The Species Group 4

QUOKKA, *Setonix brachyurus* (Plate 9). South-western Australia (including Rottnest Island and Bald Island); on mainland—dense vegetation in swamps amidst dry sclerophyll forest; on islands—anywhere.
Recognition: size rather larger than domestic cat; tail uniform colour, and sparsely haired; face very short, ears rounded and set on top of head.

Group 5   Nail-tailed wallabies Plate 10

Three species of rather strikingly marked wallabies are called nail-tailed wallabies because they possess a small dark horny nail, rather like a finger nail, hidden in the dark hair at the end of the slender tail. It is, however, not the only feature that they have in common; their incisors and premolar teeth have not the usual shape found in wallabies and this unusual dentition, together with their characteristic coat patterns, argue that they are a united group of related species.

The nail-tailed wallabies are very rare nowadays except in the far north where the Karrabul, or Nothern Nail-tail, is quite common. The three species are similar, being brightly marked with white cheek-stripes, hip-stripes, and a prominent stripe running from behind the arm up on to the back. In the Karrabul, these markings are less prominent than they are in the others and the shoulder marking less extensive.

Nail-tailed wallabies have also a clearly defined stripe along the middle of the back and the tail is long and slender. The Karrabul is the largest of them attaining a total length from nose to tail of about four feet and a weight of about twelve pounds. The roman nose, square muzzle, whip-like tail and back-stripe shown in the animal in the accompanying plate are characteristic of the group.

At one time the Wurrung, or Crescent Nail-tailed Wallaby, extended into south-western Australia but nowadays it is confined to the Centre. It is a smaller animal than the Karrabul reaching only between two and three feet from nose-tip to tail-tip. The Bridle Nail-tailed Wallaby, or Merrin, was once the most common of all small wallabies on the lower Murray and Darling; there has been no reliable record of it in New South Wales for thirty years (see p. 207).

Little is known about the natural history of nail-tailed wallabies beyond the remarks of those who collected them for museums in the early days. John Gilbert, who collected both species for John Gould, noted that the Merrin was common in thick patches of brigalow on the Darling Downs and on low, dry stony ranges covered with shrub-like stunted trees; and that in the South West of Western Australia, the Wurrung occupied patches of thick scrub, or dense thickets. Both species would take refuge in hollow logs when pursued. Fortunately in the case of the Karrabul, despite a lack of present information, there is still time for a comprehensive study to be made.

The Species   Group 5

KARRABUL, NORTHERN NAIL-TAILED WALLABY, *Onychogalea unguifera* (Plate 10). Northern Australia from Dampier Land in W.A. to the Pacific coast of northern Qld; woodland savannah, shrub savannah, between rocky ranges and rivers.
Recognition: shoulder-stripe scarcely extends above armpit.

WURRUNG, CRESCENT NAIL-TAILED WALLABY, *Onychogalea lunata.* South-central, and south-western Australia (possibly also recorded from the junction of the Murray and Darling Rivers, Vic. and N.S.W.); woodland and steppe.
Recognition: white shoulder-stripes extend to base of neck.

MERRIN, BRIDLE NAIL-TAILED WALLABY, *Onychogalea fraenata.* Interior of N.S.W. and southern Qld.
Recognition: white shoulder-stripe extends up neck to base of ear.

Group 6   Hare-wallabies   Plate 11

When the grassy plains of the inland were being opened up, the colonists found that some wallabies behaved rather like the European Hare. They made nests under tussocks or in bushes and, when flushed from these, came out with a rush, and set off at high speed. They called them hare-wallabies.

There are five species of hare-wallaby but, except for the Eastern and Western Hare-wallabies (and possibly the unknown species from the Centre), they do not

KARRABUL, *Onychogalea unguifera*   Group 5

appear to be particularly closely related to each other. They are superficially grouped together through the hare-like habits they have in common.

With the exception of the Banded Hare-wallaby, the hare-wallabies occur as rather solitary animals in open grassy or spinifex plains with, or without, shrubs or trees. The Banded Hare-wallaby, in strong contrast with these, is a gregarious animal which congregates in the spaces under the low-hanging limbs of bushes in dense thickets. Today, only the solitary, and seldom-seen, Spectacled Hare-wallaby of the northern Centre appears to be secure on the mainland, and it is very common on Barrow Island. Although the Western Hare-wallaby and the Banded Hare-wallaby seem to have disappeared from the mainland, both occur on Bernier and Dorre Islands off the Western Australian coast. The Western Hare-wallaby may still be secure in the country about the Canning Stock Route.

It is not known why hare-wallabies are rare on the mainland today but it is usually put down to predation by the Fox, because they are rather small animals and are certainly small enough for the Fox to take without difficulty. The explanation is adequate for a small gregarious mammal which congregates under dense, over-hanging bushes, as the Banded Hare-wallaby does; such an animal would certainly be very vulnerable to introduced predators such as the Fox and the Domestic Cat gone wild. But the reason for the current rarity of the other solitary species is probably more complex; their numbers may have been reduced through other factors such as the alteration of their open grassland habitat through trampling and grazing by sheep and cattle.

The populations of the Banded Hare-wallaby and of the Western Hare-wallaby on Bernier and Dorre islands, in Shark Bay, Western Australia, are of great importance to conservationists because information on them, and other mammals of the islands (including the Boodie — a rat-kangaroo, see p. 66), has been recorded by visiting scientists at intervals during the last 170 years. In fact, these islands are valuable natural laboratories where managers of reserves will learn about the way in which populations of several closely-related species interact, and fluctuate, within a confined area.

From the information available to us it seems that the Banded Hare-wallaby has always been common in the islands since Péron visited them in 1801; nevertheless, its population fluctuates between huge numbers and relative scarcity. In 1959, when we visited the islands, Banded Hare-wallabies were moderately abundant; the population appeared to be in a healthy state and, as is usual in a stable population of a mammal species, there was evidence of heavy mortality among juveniles; follow-ing this, death occurred only relatively rarely until old age was reached. Their principal predators seem to be birds of prey, but reptiles, such as the Racehorse Goanna (*Varanus gouldii*) and Children's Python (*Liasis childreni*), also occur on the islands and would be capable of taking young wallabies as well as the mice and bandicoots which form their more usual diet.

In 1959 the Western Hare-wallaby was very rare but, when we revisited the islands in 1963, we found the Western Hare-wallaby quite common in the same places; the Banded Hare-wallaby and the Boodie did not appear to have altered in abundance since our earlier visit in 1959.

The speed and jumping powers of the solitary hare-wallabies which live out on the open plains are legendary. John Gould gave a vivid account of these in an Eastern

11   SPECTACLED HARE-WALLABY, *Lagorchestes conspicillatus*   Group 6

Hare-wallaby which was coursed by two fast dogs for half a mile and then, having been chased to within twenty feet of him, it bounded cleanly over his head.

The Species

All hare-wallabies have well-marked orange rings of hair around their eyes—a feature which is particularly obvious in the Spectacled Hare-wallaby.

SPECTACLED HARE-WALLABY, *Lagorchestes conspicillatus* (Plate 11). W.A. (southern Kimberley, Pilbara and Barrow Island), North-Central Australia (to approx. 24° South), Qld (western-central, and north-coastal); spinifex grasslands.
Recognition: very conspicuous orange ring around eye; square face; marked white hip-stripes.

WESTERN HARE-WALLABY, WURRUP, *Lagorchestes hirsutus*. Central W.A. (including islands of Shark Bay), Central Australia; spinifex grasslands.
Recognition: Orange ring around eye; long orange hair behind legs around hind quarters.

EASTERN HARE-WALLABY, BROWN HARE-WALLABY, *Lagorchestes leporides*. Western N.S.W., eastern S.A., north-western Vic. along Murray.
Recognition: orange ring around eye; black patch on elbow.

Lagorchestes asomatus. Known from solitary skull from Lake McKay, N.T.-W.A. border.
Recognition: external characters unknown.

BANDED HARE-WALLABY, MUNNING. *Lagostrophus fasciatus* (Plate 1, p. 9). South-west of W.A., islands of Shark Bay (an early record from S.A. is unsubstantiated by any specimen but the Banded Hare-wallaby has been obtained in an archaeological excavation of an Aboriginal site on the lower Murray, S.A.); under low scrub in sclerophyll woodland or heath.
Recognition: dark greyish with transverse bars across lower back; size of cat; sharply constricted muzzle.

Group 7   Rock-wallabies

Plate 12

The phrases 'cat-like' or 'monkey-like' are inadequate to describe the incredible two-footed precision of a rock-wallaby leaping from ledge to ledge up a cliff face and, as might be expected, its feet and tail possess adaptations which give it a superb grip on rocky surfaces and perfect balance; the feet have thick pads which are covered with rough granulations all over the sole and are outlined in a fringe of stiff hairs; the tail is long and cylindrical, being covered with long hair and tufted at the end in keeping with its use as a balancer.

The manner in which the physical characters of the rock-wallaby fit it for its rocky and precipitous home is given graphic and unusual emphasis in a description by Professor Wood Jones of a Brush-tailed Rock-wallaby travelling over both open country and rocks on Pearson Islands at the eastern end of the Australian Bight. There, in the security of its island home, and in the absence of competitors, it moves freely in open country. In his account Professor Wood Jones said:

12   ROTHSCHILD'S ROCK-WALLABY, *Petrogale rothschildi*   Group 7

On such level ground as the island affords its gait appears somewhat awkward, for it travels with the head low and the tail arched conspicuously upwards. It seems that for such progression it has to cant its body forwards at an ungraceful angle, for the tail is not used as a fulcrum as it is in the 'scrub' wallabies and kangaroos, it is carried sheer off the ground in all gaits. When one is started across the more-or-less level saltbush areas it gets away at an awkward gait, using every bush for cover as it goes, but seeming to proceed more-or-less without regard to its bearings. When it wishes to see where its safest line of retreat lies, or where it is threatened, it stops, puts up its head, and looks around. But short of stopping, it appears unable to raise its head from its rather ungraceful stoop to take in any wide view. Though this may seem an ungraceful animal in open bush country, it is a very different creature when seen upon the huge, fantastic, granite boulders which constitute the main portion of its island home. Here its movements are astonishing; there seems to be no leap it will not take, no chink between boulders into which it will not hurl itself. There is no part of the northern portion of the islands that it does not inhabit; it is at home on the naked granite boulders of the shore upon which the surf crashes, and on the lichen covered boulders of the summit, nearly 800 ft. above, where moss, ferns, and casuarinas of large growth constitute a very distinct type of environment.

Like all the Kangaroo family, the rock-wallabies have usually only one young at a time. They shelter among the great rocks which are their homes and come out to feed in the late afternoon or evening. However, they may be watched at other times because, secure around their rocks, they often sun themselves and move about during daylight hours. The presence of rock-wallabies can usually be detected by the high gloss which their feet give to the rocky floors of the caves in which they live, and by their droppings, which are elongate, not rounded like those of other wallabies.

No modern studies have yet been published on the behaviour and ecology of rock-wallabies but it is likely, from casual observations which have been published, such studies will be fascinating. They will reveal the ways in which rock-wallabies are adapted, both physically and physiologically, to their highly specialized environment and, since they dwell in fairly large communities and appear to be active and quarrelsome creatures which chase each other about, their behaviour patterns are likely to be of very great interest as well.

The distribution of the Brush-tailed Rock-wallaby is unusually wide for an Australian mammal; it may be that the constant temperature afforded by its rocky home, which contains deep caves and fissures under rock ledges, gives it a stable environment and permits it to occur in the wide range of climatic zones within the Australian continent. The effectiveness of this habitat in providing an even temperature has been demonstrated by Dr E. H. M. Ealey, of the Division of Wildlife Research, C.S.I.R.O., in his studies of the ecology of animals of a rocky outcrop on Woodstock Station, near Marble Bar, in the Pilbara of Western Australia. Dr Ealey found that this outcrop (which included among the species living in it a population of Rothschild's Rock-wallaby) contained deep caves between the rocks in which the air temperature did not fall below 80° F. or rise above 90° F. while air temperatures in the shade outside fluctuated daily between 65° and 115° F.

At present there is no certainty about how many species there are; however, from their physical features, it seems that among the rock-wallabies there are a number of different stages of evolution. It is almost as if we are looking at a tree in which some of the branches left the stem near to the bottom and are very different from those which diverged near to the top. Thus, among the rock-wallabies, some species (the Yellow-footed Rock-wallaby and the Little Rock-wallaby) are very different from each other and from other rock wallabies; while among the others, which are obviously very much like each other, two species (the distinctive Purple-necked Rock-wallaby and Rothschild's Rock-wallaby) are somewhat like each other but different from the remainder. The remaining rock-wallabies, although recognizably different from each other by colour, probably represent no more than different populations of a single species (the Brush-tailed Rock-wallaby) which occurs widely over much of Australia (Godman's Rock-wallaby and the Short-eared Rock-wallaby are probably closely related to this species). The Brush-tailed Rock-wallaby is the species familiar to visitors to Jenolan Caves in New South Wales, Cunningham's Gap and Queen Mary's Falls in Queensland or the Macdonnell Ranges near Alice Springs. In the western, central, and northern parts of its range, this rock-wallaby is called the Western Rock-wallaby; in north Queensland—the Pale Rock-wallaby; and on the islands off the southern coast—the Pearson Islands Rock-wallaby and the Recherche (or Hackett's) Rock-wallaby.

The Species

Group 7

All live among the boulders of rock piles, rocky hills, and cliffs.

BRUSH-TAILED ROCK-WALLABY (including Western Rock-wallaby, and Pale Rock-wallaby), *Petrogale penicillata*, probably all over Australia. Not Tas.
Recognition: colour pattern striking in south, less distinct in north-eastern part of range. In southern and western districts, black cylindrical bushy tail, black armpit with pale stripe behind it, vivid white cheek-stripe, ears with prominent black patch and whitish margins; in northern parts of Queensland stripes and patches may be scarcely noticeable. In all, tail cylindrical, darker at end.

ROTHSCHILD'S ROCK-WALLABY, *Petrogale rothschildi* (Plate 12). Central-western W.A. (Hamersley Range, Pilbara district, Dampier Archipelago).
Recognition: no striking pattern except for brilliant purple upper back at certain times of year, fur yellowish; may be distinguished from *P. penicillata* in its range where *penicillata* has pale stripe behind black armpit and bicoloured black and whitish ears.

PURPLE-NECKED ROCK-WALLABY, *Petrogale purpureicollis*. North-western Qld (vicinity of Mount Isa).
Recognition: no striking pattern except for brilliant purple upper back at certain times of year.

GODMAN'S ROCK-WALLABY, *Petrogale godmani*. Cape York Peninsula (vicinity of Cooktown).
Recognition: terminal part of tail pale, part closest to rump darker.

SHORT-EARED ROCK-WALLABY, *Petrogale brachyotis*. Northern Kimberley, northern N.T. including Arnhem Land.
Recognition: no striking pattern but clearly a rock-wallaby; within its range the brightly coloured forms of *penicillata* and *Peradorcas concinna* can easily be distinguished from it by their striking ornamentation.

YELLOW-FOOTED ROCK-WALLABY, RING-TAILED ROCK-WALLABY, *Petrogale xanthopus*. South Australia (Gawler Range, Flinders Range), north-western New South Wales (and around Broken Hill), south-western Qld (Main Barrier Range, Grey Range).
Recognition: tail with rings (indistinct in south-western Qld); hind feet and fore arms yellow; very large furry ears, yellow and white.

LITTLE ROCK-WALLABY, *Peradorcas concinna*. Northern Kimberley, northern N.T. including
Arnhem Land. Sandstone country including rocky cliffs and hills.
Recognition: very small, cat size; upper back rusty-red to greyish, rump reddish to orange,
tail darker towards end.

Group 8   Tree-kangaroos   Plate 13

Among the mammals which crossed between Australia and New Guinea, during
the times that Torres Strait was exposed as land, are two species of tree-kangaroo.
Both of these are so different from their New Guinea relatives that they are usually
regarded by zoologists as fully differentiated species. One of these, Lumholtz's Tree-
kangaroo, was discovered by the Norwegian naturalist, Dr Carl Lumholtz, while the
other, Bennett's Tree-kangaroo, was named after the distinguished medical man and
naturalist, Dr George Bennett, who was one of the founders of the Australian
Museum.

Both species live in the mountainous rain forests of the eastern side of north-eastern
Queensland. As their name suggests, they are kangaroos which have forsaken the
terrestrial life of their relatives and have taken up tree-dwelling. In response to the
requirements of their arboreal habitat, tree-kangaroos have short, broad hind feet
and larger and better developed fore arms than other kangaroos. Their tails are
cylindrical and rather rock-wallaby-like.

Despite the fact that they are agile tree-dwellers, they are said to spend a great deal
of their time on the ground but climb back into the trees when alarmed or pursued.

The Species   Group 8

LUMHOLTZ'S TREE-KANGAROO, BOONGARY, *Dendrolagus lumholtzi*. North-eastern Qld
(around Cairns, south of Daintree River); forest trees.
Recognition: powerful fore arms, short broad hind feet, long cylindrical tail, ears short
and rounded; face dark with pale forehead; hands and feet dark contrasting with grey
back and pale whitish belly.

BENNETT'S TREE-KANGAROO, DUSKY TREE-KANGAROO, TCHARIBEENA, *Dendrolagus bennettianus*. North-eastern Qld (around Cooktown, north of Daintree River); forest trees
on or near tops of mountain ridges from 1,500 to 2,500 feet.
Recognition: Powerful fore arms, short, broad hind feet, long cylindrical tail; ears short
and rounded; general colour of head and body brown.

Group 9   Rat-kangaroos   Plate 14

There are five divisions, or genera, of rat-kangaroos; one of these, the small
Musk Rat-kangaroo, *Hypsiprymnodon*, of Queensland is so different from the others
that it is often put in a separate sub-family. The others fall into two more or less

13 TREE-KANGAROO (a New Guinea species), *Dendrolagus goodfellowi*   Group 8

distinct kinds, the bettongs and their relatives, and the potoroos. The bettongs are mammals of fairly dry country. One species (the Boodie) digs burrows, but the rest make nests of grass on the surface. Bettongs carry their nesting material curled up in their tails, a habit which they share with the Rufous Rat-kangaroo. Bettongs are partly carnivorous and in captivity are avid meat-eaters; they also gnaw bones.

The bettong group includes also the Plains or Desert Rat-kangaroo, *Caloprymnus*, which, as its common name suggests, is an animal of the desert plains of Central Australia. This animal was 'missing' for many years and was rediscovered, but now seems 'lost' again (see page 198). The Rufous Rat-kangaroo is also a member of the Bettong group. It makes its nest in dense grasses in the tall woodland of north-eastern New South Wales, and in eastern Queensland as far north as Cairns. The Rufous Rat-kangaroo is one of the many species of mammals which appears to have been declining in geographical range prior to European occupation, possibly due to some long-term climatic fluctuation, but it may now have reached a stable distribution (see page 5).

The remaining group of rat-kangaroos are the potoroos which at a casual glance look rather like bandicoots. They are creatures of tussock and densely growing grasses associated with forest, or woodlands, or other dense vegetation such as heaths in areas of high rainfall. Studies which were made near Smithton, Tasmania, by Dr George Heinsohn show that the Potoroos apparently require the dense natural vegetation of their habitat for protection against their many potential predators which, since they are small mammals, include birds of prey, reptiles and native carnivorous mammals as well as introduced predators. They are strictly nocturnal in their behaviour and, unlike wallabies and bandicoots living in the same area, they stay entirely within their dense cover and do not come out of it to feed in adjacent open pastures. If the habitat is opened up either by fire or clearing Potoroos disappear until the vegetation has recovered. Such species are vulnerable to burning-off in bushfire control.

Dr Heinsohn's work suggests that female Potoroos may have joint or overlapping home ranges whereas males occupy separate territories. This is supported by observations made by Dr R. L. Hughes, who found that females kept together in captivity did not fight with each other but severe fighting occurred between males when they were together with a female which was on heat. Dr Heinsohn says that none of the animals which he observed in his work on the wild populations had wounds or scars; it thus seems likely that their behaviour patterns are such that, in the natural situation, they avoid coming into conflict.

Potoroos appear to be quite long-lived for small mammals; Dr Eric Guiler and Mr D. Kitchener have established that, in Tasmania, Potoroos may live for as long as seven years in the wild. Young are not weaned for at least twenty-one weeks after birth, but they come out of the pouch in their seventeenth week, and from the fact that Dr Heinsohn caught one at about this age, it appears that they are running free at about this time.

The Potoroo breeds throughout the year in Tasmania. Each female seems to commence breeding when it is about one year old, and is also able to breed twice in each year, but the reproductive potential of the Potoroo is low because, like other members of the kangaroo family, it only produces one young at a time. Mating follows about four days after the birth of the first young, but the resulting embryo

14 BRUSH-TAILED BETTONG, *Bettongia penicillata* Group 9

delays development for as long as four and a half months until either the first young is suckling intermittently during its weaning period or it is prematurely lost from the pouch (see p. 42 for an account of the delayed birth in kangaroos).

The Musk Rat-kangaroo of the Queensland rain forests, one of the least-known members of the Australian mammal fauna, is the exception to the rule that members of the kangaroo family have only one young at a time; this little rat-kangaroo produces two.

Of the species of bettong, the Boodie has become best known as a result of studies which have been made at the Australian National University and in the Division of Wildlife Research, C.S.I.R.O., at Canberra, upon a captive population which has been bred from stock obtained on one of our visits to Bernier Island, in Shark Bay, Western Australia. The behaviour of this population was studied by Dr Eleanor Stodart in a large enclosure which had formerly housed a population of rabbits; the Boodies converted the rabbit burrows to their own use. They changed the outward appearance of the burrows by digging out the entrances with their fore paws into long open channels which led to the mouths of the burrows. These channels are an easily recognizable feature of the burrows of Boodies in the wild. Despite the fact that Boodies make only simple nests containing little vegetation (unlike their surface-dwelling relatives) they expend much activity on nest-building; they collect straw in their mouths, place it on the ground between their hind feet, use the hind feet to push the hay back over the downwardly curled tip of the tail, tramp it down and then carry it as a bundle within the tail. They are entirely bi-pedal in locomotion; they never use their fore feet, even when moving slowly. Moreover, the tail is not used for support as it is in the kangaroos and most wallabies.

Boodies are entirely nocturnal; they appear above ground after sunset and disappear well before sunrise. The males emerge first, followed by the females, and are shy and easily frightened until it becomes quite dark when they become confident, and some will even approach intruding humans. Males start their evening's activities by exploring the whole of the enclosure and investigating each female. From Dr Stodart's work it seems that Boodies form social groups in which a male defends its group of females rather than any particular territory; although females will establish territories and exclude other females from them. Males are aggressive towards each other but females are generally amicable and communal in behaviour. During the day the males usually lie up with a female but females may rest together. Females are sexually attractive to males well beyond the periods that they are receptive and attempts to mate are frequent. Dr Hugh Tyndale-Biscoe has studied breeding in this colony and has shown that birth follows on the twenty-second day after mating. A further copulation follows birth and the resulting embryo is delayed in its development for as long as four months while the previous joey is suckled in the pouch. Young leave the pouch permanently when they are between 113 and 120 days old. From animals which were reared in captivity from the Bernier Island population it seems that females are capable of giving birth when they are about two hundred days old and a female brought back from Bernier Island by Hugh Tyndale-Biscoe and me in 1959 reared three young to independence in one year. Professor T. T. Flynn has observed that females of the Tasmanian Rat-kangaroo will mate soon after they are free of the pouch.

The information that is available suggests that Rat-kangaroos in general produce young throughout the year, but Dr Tyndale-Biscoe's work suggests that, in the Bernier Island population of the Boodie, at any rate, there is a seasonal breeding pattern. Most births occur between February and September but, although other births occur later in the year, these may be the result of delayed embryos from pre-September matings. In the captive stock most females did not ovulate during November and December although some mating occurred during that time.

Boodies fight by kicking, lying on their sides lashing out with their hind feet, but when held by hand will also bite savagely.

We have a long way to go yet before we understand fully the arrangement of the species of rat-kangaroos. In recent years, studies by Mr H. H. Finlayson and Mr Norman Wakefield have helped us to a greater knowledge of classification of the bettongs; while the potoroos have recently formed the complex subject of analysis by computer by Mrs Jeanette Hope of Monash University. The results of her work show that there are distinct populations which can be mathematically defined. These populations live in parts of south-eastern Australia, in the Bass Strait islands, in Tasmania, and in south-western Australia. Although they are measurably different from each other, when numbers of characters are considered together the populations fall into two wider groups which probably represent two distinct species. One group contains the populations centred on New South Wales and south-western Australia, and the other contains the rest. This work is the basis of the classification of potoroos given here.

The Species

Group 9

All but the Rufous Rat-kangaroo are small—the size of cats or smaller.

MUSK RAT-KANGAROO, *Hypsiprymnodon moschatus*. North-east Qld; rain forest on the eastern side of the mountains from Mossman to Townsville.
Recognition: five toes on hind foot, tail naked, size of rabbit.

DESERT RAT-KANGAROO, PLAINS RAT-KANGAROO, *Caloprymnus campestris*. Extreme north-east of S.A., extreme south-west of Qld, open plains and sandridge desert.
Recognition: general colour pale; tail well-haired below, often naked on upper surface; head blunt, rounded; front limbs very small, 1/3 of very long hind limbs.

RUFOUS RAT-KANGAROO, *Aepyprymnus rufescens*. North-eastern Vic., eastern N.S.W. (north of Newcastle), and eastern Qld; sclerophyll woodland and forest with dense grasses.

Recognition: hair on snout extending between nostrils; general colour greyish with reddish-chestnut overlay, paler below; distinct white hip-stripe; hair rather long with coarse texture; ears very hairy.

WOYLIE, BRUSH-TAILED BETTONG, BRUSH-TAILED RAT-KANGAROO, *Bettongia penicillata* (Plate 14). South-western Australia, southern S.A., north-western Vic. (Murray), central N.S.W.; sclerophyll woodland.
Recognition: tail with long hair at tip and for some inches back forming black crest.

EASTERN BETTONG, TASMANIAN RAT-KANGAROO, GAIMARD'S RAT-KANGAROO, *Bettongia gaimardi*. South-eastern Qld (coastal), N.S.W. (coastal), southern to south-western Vic., Tas.
Recognition: tail with long hair forming crest as in *B. penicillata* but last $\frac{1}{2}$ inch to $1\frac{1}{2}$ inches almost invariably white.

NORTHERN RAT-KANGAROO, *Bettongia tropica.* Eastern Qld.
Recognition: crested tail with or without silvery white tip; not distinguishable from *B. penicillata* or *B. gaimardi* externally.

BOODIE, TUNGOO, LESUEUR'S RAT-KANGAROO, BURROWING RAT-KANGAROO, *Bettongia lesueur.* W.A. (Dampier Land, islands off the west coast, south-western and central western), S.A., N.T. (south of 20°), south-western N.S.W.; constructs burrows in steppe, heath, or sclerophyll woodland.
Recognition: tail not long-haired (tends to be fat-tailed), tip white in most populations; burrows.

POTOROO, LONG-NOSED RAT-KANGAROO, GILBERT'S POTOROO, *Potorous tridactylus.* South-eastern Qld, coastal N.S.W., north-eastern Vic., south-western corner of W.A.; dense grasses and low thick scrub, particularly in damp places.
Recognition: pointed bandicoot-like head, naked patch extending up on to snout from nose; often but not always white tip to tail; hops with fore part of body close to ground.

POTOROO, SOUTHERN POTOROO, *Potorous apicalis.* Eastern and southern Vic., S.A. (probably south-eastern only), Tas., islands in Bass Strait; dense grasses and low thick scrub, particularly in damp places.
Recognition: pointed bandicoot-like head; externally similar to *P. tridactylus*; white tip to tail almost always present.

BROAD-FACED POTOROO, *Potorous platyops.* South-west of W.A. (inland periphery from Pallinup R. to Goomalling)*.
Recognition: short head; very small, about size of rabbit.

---

*Formerly occurred in Kangaroo Island, S.A., and on the lower Murray where its remains are found in superficial cave deposits and in Aboriginal campsites. Not recorded in these places in European times.

6

Possums, the Koala and wombats

THE POSSUMS, the Koala, and the wombats, together with the kangaroos, make up the living members of the large branch of marsupials which are properly called the Diprotodonta. In the Pleistocene period of the last two million years the group was even more diverse than it is today and included such forms as the *Thylacoleo*, or Marsupial Lion, the heavily built *Diprotodon* and *Nototherium* and their more slenderly built relatives, the palorchestines. Like modern wombats, these large animals were terrestrial and, when on all fours, the largest of them stood up to five feet high at the shoulder.

If evolutionary success can be measured by the number of surviving genera, the Diprotodonta are the most successful marsupials in Australia. Today they have thirty-three genera—the native cat family has twelve, and the bandicoots only five.

Among the Diprotodonta is the earliest, certainly dated, Australian marsupial. This is *Wynyardia bassiana* from the Oligocene or lower Miocene beds of Table Cape, near Wynyard, in northern Tasmania. It seems to have been a possum-like creature which lived some thirty million years ago and it possesses various primitive features which indicate its descent, and that of the other Diprotodonta, from the native-cats and their allies.

Possums, the Koala and wombats are all diprotodont and syndactyl (see p. 39). Unlike the Koala and the wombats, the possums have forward-opening pouches like kangaroos. They are mostly tree-dwellers who are less specialized than the kangaroos for locomotion on the ground; their hind feet are invariably hand-like and obviously very different from the elongate feet of the kangaroo and bandicoot families. Even the most possum-like of the kangaroo family, the potoroos and the Musk Rat-kangaroo, can easily be told apart from any possum by this means.

A walk through native bush at night with a powerful torch will show you that possums and their relatives are quite common animals. They are mostly arboreal but many of them spend a good deal of their time on the ground crossing from one tree to another or even feeding.

The Koala and the wombats belong to a separate evolutionary branch of marsupials from the rest of the possums and the kangaroos; and are very different from even the largest of ordinary possums in general appearance. For many years the Koala was thought to be a sort of dumpy ringtail, but it is now generally recognized that the numbers of specialized characters of anatomy, blood proteins, and chromosomes which it has in common with the wombats indicates true relationship with them at about the same degree to which the possum family is related to the kangaroo family. The Koala and the wombats differ from possums and kangaroos, and resemble bandicoots and the Thylacine in that the pouch opens backwards. One can appreciate that this feature could be a useful adaptation in a burrower like a wombat; but it is more difficult to understand in a climbing animal like the Koala.

Group 10 Large possums and cuscuses   Plates 3, 15, 16

Most Australians dwell in suburbs and, under these conditions, they have more contact with the common Brush-tailed Possum (sometimes called the Silver Grey) than they have with any other native mammal. There, it hides in roofs and eats cultivated plants and blossoms. But it is also a familiar animal to the bushman, because it is one of the most widely occurring mammals in Australia.

There are three fairly closely related genera of large possums. These are the common possums, the Scaly-tailed Possum of the Kimberley, and the cuscuses of Cape York; they are treated here as a single group. The smaller possums, the flying possums and ringtails, are dealt with separately.

As far as is known, the female Brush Possum starts breeding at the end of its first year, and it is probable that on the mainland she only gives birth to one young each year of her reproductive life; in Tasmania, Dr Eric Guiler believes that females breed twice each year. One young is the normal number, but twins, and even triplets have been recorded. The joey leaves the pouch after five months.

The Brush Possum is a relatively sedentary and solitary animal and Dr G. M. Dunnet of the C.S.I.R.O. found that, near Canberra, males and females have discrete and limited ranges. Females occupy about $2\frac{1}{2}$ acres while males occupy as much as $7\frac{1}{2}$ acres; male territories overlap with those of several females.

The Brush Possum has a variety of marked colour forms and this has led biologists to describe numbers of species. It is probable, however, that there are only three species of the genus *Trichosurus* in Australia (while in the other genera there is one of *Wyulda*, and there are two of *Phalanger*). Among the most prominent colour phases of the Brush Possum is a black one which is particularly common in Tasmania. Dr Eric Guiler has shown that this phase appears to be dependent upon climate in its distribution and it is found predominantly in areas of thick sclerophyll forest where there is an annual rainfall of more than forty inches. In all colour phases of the Brush Possum the large ears, prominent pink nose and pink finger-like end to the tail are conspicuous features which enable the species to be recognized.

Brush Possums are economically important. In Tasmania the Brush Possum is said to be responsible for considerable damage in regenerating eucalypt forests, and in Victoria it and the Bobuck are destructive in plantations of introduced pine trees. The Brush Possum is also an important fur-bearer, particularly in Tasmania, where the annual crop is of considerable value. There, from 1923 to 1955 a total of 1,062,513 skins were marketed; numbers taken in recent years include 90,815 in 1949 and 61,357 in 1952. In Victoria 107,500 were taken during the 1959 season.

The Scaly-tailed Possum of the Kimberley is one of our least-known animals (see page 18). It feeds on blossom and appears to spend much time among the rocks in broken sandstone country. Harry Butler, who searched successfully for it on behalf of the American Museum of Natural History and the Western Australian Museum, says that his observations suggest that it is solitary, like the Brush Possum. It carries one young at a time and, from the three young specimens known, one can only say of breeding that naked pouched young are carried in West Kimberley in June, and that by December and January in North Kimberley they are about half grown.

The cuscuses are strange looking round-headed almost earless creatures with tails

15   BRUSH POSSUM, *Trichosurus vulpecula*   Group 10

which are bare from about half-way down. There are two Australian species which are confined to Cape York; they have obviously invaded Australia from New Guinea and the islands to the north, where they are common and widespread from Celebes in the west to the Solomon Islands in the east.

The Species   Group 10

BRUSH-TAILED POSSUM, BRUSH POSSUM, BUSHY-TAIL POSSUM, COMMON POSSUM, SILVER GREY, *Trichosurus vulpecula* (Plate 15). All Australia except the top end of the N.T., the Kimberley and Barrow Island; usually in trees, but where these are few, will utilize caves and holes in the ground.
Recognition: as in Plate (but white tail-tip only common in W.A.); general colour variable from black to grey; about size of cat.

NORTHERN BRUSH POSSUM, *Trichosurus arnhemensis.* Kimberley, top end of N.T., Barrow Island; usually in trees (Barrow Island, mangroves, holes in limestone).
Recognition: more slender, tail less well-haired than in Plate and looks thin compared with that of *vulpecula*; general colour grey.

BOBUCK, MOUNTAIN POSSUM, SHORT-EARED POSSUM, *Trichosurus caninus.* Mountain districts of south-eastern Qld, eastern N.S.W., eastern Vic.; forest.
Recognition: tail bushy close to body but shorter hairs towards tip causes pronounced taper; ears short and rounded; general colour dark grey or black with glossy sheen.

SCALY-TAILED POSSUM, *Wyulda squamicaudata* (Plate 3, p. 19). Western, central, and northern Kimberley; trees in rocky country.
Recognition: tail thickly furred at base then suddenly naked with scales giving rasp-like appearance; short close fur; general colour grey to brownish-grey.

SPOTTED CUSCUS, *Phalanger maculatus* (Plate 16). North-eastern Cape York from McIlwraith Range northwards; rain forest and gallery forest.
Recognition: terminal half of tail naked, head round, ears very short, no dorsal stripe, general colour blotched or plain, often very pale.

GREY CUSCUS, *Phalanger orientalis.* Eastern Cape York from McIlwraith Range northwards; rain forest and gallery forest.
Recognition: Terminal half of tail naked, head round, ears very short, back grey or grey-brown with dark dorsal stripe.

Group 11 Ringtails   Plate 17

Ringtails are the nest-builders responsible for the untidy looking balls of twigs and leaves, about a foot across, which are so familiar in many kinds of dense scrub in eastern and south-eastern Australia. These nests, or dreys, are particularly common among the coastal Ti-tree scrub (*Leptospermum*) around the shores of Port Phillip Bay, south of Melbourne. Ringtails do not only occur in coastal scrubs, for they favour sites wherever trees are closely spaced together as they are around swamps; as a result they range from the coastal scrub to the highlands. In Victoria they even go

16　SPOTTED CUSCUS, *Phalanger maculatus*　　　　　　　　　　　Group 10

as high as the edges of Snow Gum country. They are less common in Western Australia where they are confined to the South West and are most often to be found among peppermint trees (*Agonis*). As the name ringtail implies, the tail is strongly prehensile.

There are a number of species of ringtails which are usually regarded as belonging to three genera. The Common Ringtail and other 'true ringtails' are called *Pseudocheirus*; the Brushy-tailed Ringtail of the rain forests of the Atherton Tableland is called *Hemibelideus*; and the Rock-haunting Ringtail of the Northern Territory and east Kimberley is called *Petropseudes*; this, unlike the rest, has a short, apparently non-prehensile, tail. Among the true ringtails is one of the most strangely coloured of all Australian mammals, the Green Possum. This ringtail has peculiar golden-green fur, a hue which Hobart Van Deusen, of the American Museum of Natural History, has shown to be produced by an unusual combination of pigment and structural effect which, by combining black, white and yellow, gives a green appearance. This possum is closely related to the green Mountain Ringtail from New Guinea.

Although it is difficult to believe it from its external appearance, the other close relative of the ringtails is the Dusky Glider or Greater Gliding Possum, that strange shaggy-furred, black and white (or sometimes pure white) flying possum of the tall gum forests of the eastern mountains.

The breeding season of the Common Ringtail is different in different parts of eastern Australia. In Victoria they have only one litter per year (exceptionally twice) and they breed in winter; in Sydney, they breed twice yearly—in early winter and in early summer; in northern Queensland, in late summer and autumn. Studies by Dr Michael Marsh, of ringtails around Sydney, reveal that forty-five to seventy-five per cent of all young possums die before they become reproductively mature at the age of eighteen months. A ringtail usually has a litter of two and, at that rate, for the population to be stable, each female surviving to breed must produce from six to eight young during her life; presumably in the wild a female Common Ringtail lives about three and a half years. The maximum life span in captivity is something more than five years.

Ringtails are nocturnal. They appear to be strictly herbivorous, eating leaves, but fruits such as the introduced blackberry also seem to form an important part of the diet in Victoria. The social behaviour of the Common Ringtail has been studied both by Dr Marsh, in New South Wales, and by Dr J. A. Thomson and Dr W. H. Owen in Victoria. From this work it seems that population density may have considerable effects on behaviour: where numbers are high, as they were in the populations studied by Dr Marsh, ringtails are very aggressive towards each other; otherwise, as in the Victorian population, individuals seem to be semi-social in their behaviour (see p. 78). Dr Marsh recorded a most unusual instance of a Common Ringtail attacking a man in the wild. The man was holding its young.

On the continent of Australia, the Ringtail is not regarded as an important fur-bearer because of the thinness of its skin; in Tasmania, however, presumably because of the colder winters, the skin is of better quality and approximately $7\frac{1}{2}$ million ringtails were taken for their fur between 1923 and 1955. By comparison, in Victoria in 1959 (the year in which over 107,000 Common Possums were taken) only 2,500 ringtail skins were marketed, most of them being taken incidentally to the much more important Common Possum crop.

Ringtails are not as large as the larger possums; they are mostly about two feet in

17   COMMON RINGTAIL, *Pseudocheirus peregrinus*   Group 11

length and smaller than domestic cats. It is probable that most of the numerous species described, including the Western Ringtail and the red Bunya Ringtail, are merely local races. Most authors treat the Tasmanian Ringtail (*Pseudocheirus convolutor*) as separate, but recent studies of their blood proteins by Dr J. A. W. Kirsch, convince me that it is merely another isolated population of the Common Ringtail.

The Species   Group 11

COMMON RINGTAIL, *Pseudocheirus peregrinus* (Plate 17). Eastern Australia from Cape York to south-eastern S.A., south-western Australia, Tas.; in eastern Australia inland from coast to western slopes of dividing range in rain forest, sclerophyll forest, or woodland, particularly along watercourses or around swamps; in south-western Australia in thick vegetation in sclerophyll forest or woodland, especially in Peppermint (*Agonis*). *Recognition:* as in Plate but colour varies with locality, tail-tip usually white.

MONGAN, HERBERT RIVER RINGTAIL, *Pseudocheirus herbertensis.* North-eastern Qld (from Townsville to Cooktown); rain forest. *Recognition:* very dark brown with white under-surface; tip of tail white.

GREEN RINGTAIL, STRIPED RINGTAIL, TOOLAH, *Pseudocheirus archeri.* North-eastern Qld (Cairns–Atherton Tableland–Herbert River area); rain forests and radiating gallery forests. *Recognition:* golden green; two stripes running down back.

BRUSHY-TAILED RINGTAIL, BRUSH-TIPPED RINGTAIL, *Hemibelideus lemuroides.* North-eastern Qld (Cairns district); rain forest. *Recognition:* general appearance like ringtail but tail bushy with only small bare exposed section at end.

ROCK-HAUNTING RINGTAIL, WOGOIT, *Petropseudes dahli.* Western Arnhem Land, east Kimberley; rock piles in woodland savannah. *Recognition:* short thick tail; dark stripe along back.

Group 12 Gliders   Plate 18

The possum family is the only one of the Australian marsupials which has evolved gliding forms. Even more interesting, though, gliding specializations have appeared in at least three separate lines of possums. Thus, each of the genera of gliding possums is more closely related to a non-gliding possum than it is to the other gliders. For instance, the Feathertail, or Pigmy Glider, is a gliding pigmy possum; the Greater Glider is related to the ringtails; and the Sugar Glider to Leadbeater's Possum.

Gliders are creatures of the tree tops which glide by means of a flying membrane—an extension of their body skin. This loose skin along their sides is spread out by extending the legs so that it becomes stretched between the outer side of the hand and the ankle. The Greater Glider differs slightly from the others in that the membrane stretches from elbow to ankle and it glides with its arms bent inwards from the elbow. In all gliders, the tail is long and well furred so that it can be used as a balancer, a feature most highly specialized in the Feathertail, where the hairs grow along the sides like the vanes of a feather (see the figure on a one-cent piece).

18 SUGAR GLIDER, *Petaurus breviceps*

Group 12

The gliding abilities of the two largest gliders—the Greater and Fluffy Gliders—have been compared recently, by Norman Wakefield, under similar conditions. He found that when animals were launching themselves from high trees, Greater Gliders would glide to a predetermined landing spot up to some 175 feet away at an angle of descent of about forty degrees to the horizontal while the Fluffy Glider descended at a flatter angle (of about thirty degrees) and went much farther (about 370 feet). When the height of the launching spot and the distance to be covered required it, both species could steepen their glides; in addition they could also change direction laterally during the glide, although the Fluffy Glider appears to be more versatile in this respect than the Greater Glider. The Fluffy Glider calls loudly while it glides but the Greater Glider is silent. Norman Wakefield considers that this vocal habit may be related to the need of individual Fluffy Gliders to maintain contact as they move through the forest. Greater Gliders are very sedentary in their habits and seem to glide much less frequently. David Fleay has recorded flights of eighty yards in the Greater Glider, 110 yards in the Fluffy Glider, and fifty yards in the Sugar Glider.

Pigmy Gliders and Sugar Gliders are feeders among blossoms, while the Greater Glider feeds only on *Eucalyptus* leaves. Greater Gliders have a single young each year which seems to be born around July or August in Victoria. The Pigmy Glider has litters of up to four, while the Sugar Glider and Squirrel Glider have a pair of young, or sometimes one. All usually live in hollow spouts or holes, in eucalypts, making a nest of dry leaves inside the hole.

The glider about whose habits we have the most knowledge is the Sugar Glider; unlike most possums it lives in communities or family parties which, in the opinion of David Fleay, who has observed them, are the young of several seasons who have continued to live with the original pair. That this explanation of the origin of social behaviour in this possum is entirely reasonable is borne out by the semi-social behaviour of populations of the Common Ringtail investigated in Victoria, by Drs Thomson and Owen (see p. 74), where nests are often grouped closely together and females share them with each other during the day; under these conditions young stay with their mothers even during the raising of a second litter and several females, with a male and two generations of young, may even be found in a single nest.

A study of the social behaviour of communities of Sugar Gliders, in captivity, has been made by Dr T. Schultze-Westrum in Munich, Germany, who showed that scent produced by the animal plays a considerable role in organization (and hence peace-keeping) within the groups. This scent is mostly produced by glands situated between the eye and the ear, on the chest, and around the genital opening; they also use urine and saliva. Each animal has its own characteristic smell but, in addition, through a process of marking each other, by rubbing glandular secretions on to each other, the whole group becomes permeated with the scent of particular high-ranking males within it. In addition to marking each other, the members of the groups also mark objects in their territory with glandular secretions and excretions.

The use of glandular secretions and of urine by marsupials for marking objects within their home ranges and territories has been investigated in several other species, and provides interesting comparison with the Sugar Glider. For example, the solitary-living Brush Possum (see p. 70) has a chest gland and anal glands which seem to be used to discourage contact between individuals; the chest gland is rubbed against objects and marks them with a scent familiar to the animal in whose territory they are, while the secretion of the anal glands of males produces an aggressive reaction in other males and is probably a direct message to another possum to keep away. In the Common Ringtail, with its apparently more complex social behaviour than the Brush Possum, there are neither forehead nor chest glands, but anal glands are present; no territorial marking has been described and their secretion seems to have no effect upon other ringtails except that males can distinguish male from female scent; they may be protective because they are certainly distasteful to some predators such as the Powerful Owl. It seems likely that, in evolving social behaviour, the Sugar Glider has put to highly complex use structures which other possums possess but may use for less specialized social purposes.

Dr Schultze-Westrum says that members of a community of Sugar Gliders know their places and seldom fight among themselves (beyond some minor bickering over food, and sometimes when a female is on heat) but when members of the same sex belonging to different groups are placed in contact they will fight. Sugar Gliders are well known to be fighters in the wild, and Fleay tells of a Sugar Glider, which he released into a wild population, being killed by other gliders after its release. Several observers have given accounts of encounters between Sugar Gliders and the Tuan (see p. 110), a reputedly savage predator, in which the Tuan avoided the glider.

Like most possums Sugar Gliders are vocally expressive and Fleay distinguishes several principal calls: first, he describes, as a curiosity or warning call, a shrill yapping grunt or bark delivered while the animal remains perfectly still; secondly, he says that members of the group make quiet sibilant cries to each other; and, thirdly, that they have a cry of anger which is a sharp, droning scream which starts at full volume and rapidly runs down to a few grunts. Fleay says that the angry cry is an inseparable accompaniment to biting.

Although Sugar Gliders usually nest in hollow limbs and trunks they will also make use of convenient situations, as they present themselves, such as the deserted dreys of ringtails. Within the hollow, the nest is a large mass of leaves which during building they carry curled in their tails (rather as bettongs and ringtails do); they bite off the leaves and place them in their tails with their hands.

Their two young stay in the pouch for about two months after birth but then, being too large to be carried within the pouch, they are left in the nest when the group are off feeding. When they first leave the pouch their eyes are closed but they open soon afterwards. It seems that it takes a while for the babies to learn the odour of the different females in the family group because for the first five days they will crawl into other pouches and suckle for short periods. At about four months they begin to move about outside the nest as part of the family party.

The closely-related Squirrel Glider is much like the Sugar Glider in its appearance and, as far as they are known, in its habits. The two species will interbreed in captivity but it is not known if they ever do so under natural conditions. Fleay has said that the Squirrel Glider does not make the curiosity, or warning, call which he recognized in the Sugar Glider.

As well as eating blossoms, both species feed on the sap of branches which they strip and pierce with their powerful incisors; they also feed on insects and their larvae. According to David Fleay, Squirrel Gliders will readily take small birds, and mice, too, and he describes an occasion on which Squirrel Gliders in captivity even attacked a half-grown Guinea Fowl which was kept in the same pen with them.

The Species   Group 12

All gliders have a flap-like extension of their body skin which can be stretched between fore and hind limbs.

PIGMY GLIDER, FEATHERTAIL GLIDER, FLYING MOUSE, *Acrobates pygmaeus*. Eastern Qld to south-eastern S.A., extending inland across the Dividing Range into the plains as far west as Deniliquin (N.S.W.); sclerophyll forest and woodland.
Recognition: hair of tail arranged like feather, mouse-sized.

SUGAR GLIDER, *Petaurus breviceps* (Plate 18). From south-eastern S.A. to Cape York, Tas. (introduced), northern N.T., north Kimberley; sclerophyll forest and woodland.
Recognition: flying membrane from wrist to ankle; dark stripe from face between eyes to mid-back; fur of back fine velvety fawn-grey to blue-grey, belly pale grey to medium grey. Total length about fifteen inches.

SQUIRREL GLIDER, *Petaurus norfolcensis*. Eastern Vic., eastern N.S.W., eastern Qld, (extends farther inland than the Sugar Glider); sclerophyll forest and woodland.
Recognition: not easily separable from *P. breviceps* on external features except size a little larger, total length about twenty inches; belly colour white or creamy white, sometimes with merest pale grey tinge; in specimens from the same locality dorsal ·stripe more intense in *norfolcensis*.

FLUFFY GLIDER, YELLOW BELLIED GLIDER, *Petaurus australis*. Coastal Qld, N.S.W. and Vic. (mainly mountain country); sclerophyll forest.
Recognition: flying membrane from wrist to ankle; fur of back grey-brown; belly often yellow or orange; hands and feet black; total length about thirty inches.

GREATER GLIDER, DUSKY GLIDER, *Schoinobates volans*. Eastern Australia from Dandenong Ranges to Rockhampton, Qld; sclerophyll forest and tall woodland.
Recognition: shaggy fur; flying membrane from elbow to ankle; nearly cat-sized; under-surface from chin to vent white, upper surface in normal form very dark grey to black but may be creamy white, pale grey, or mottled with grey, or normal with white head.

Group 13 Leadbeater's possum   Plate 19

Until recently regarded as extinct, Leadbeater's Possum was rediscovered in 1961 in Victoria (see page 12). Nowadays it is known from some four hundred square miles at heights between 2,400 and 4,000 feet in the ranges between Lake Mountain, east of Marysville and Tanjil Bren east-north-east of Noojee, far distant from the places in the south-east from which it was obtained in the early days. The presence of its remains in the cave deposits of eastern Victoria and in eastern New South Wales suggests that the species may yet be found through the Australian Alps across the Victorian border. Most of the recent sightings have been by members of the Mammal Survey Group of the Field Naturalists' Club of Victoria.

Leadbeater's Possum is believed to mate about September; young have been found in the pouch in October and November. The usual number of young is thought to be two, but this is not known for certain; there are four teats in the pouch. The nest is made of loosely matted bark placed in a hollow tree.

The feeding habits of Leadbeater's Possum in nature are known only from the stomach contents of a specimen which was collected soon after it had fed. This

19   LEADBEATER'S POSSUM, *Gymnobelideus leadbeateri*   Group 13

individual had been feeding on beetles and crickets; it is probable that they eat other insects as well. No vegetable matter was found in the stomach.

The Species   Group 13

LEADBEATER'S POSSUM, FAIRY POSSUM, BASS RIVER POSSUM, *Gymnobelideus leadbeateri* (Plate 19). Southern and south-eastern Vic.; dense wet sclerophyll forest.
Recognition: grey with stripe along back from head to mid-back; tail long, well haired, and to the tail-tip the same colour as the back (i.e. it is not black); no flying membrane.

Group 14 The Striped possum   Plate 20

Like the Green Ringtail, the cuscuses and the tree-kangaroos, the Striped Possum is probably one of the invaders of Cape York from New Guinea where its relatives are found today. The Striped Possum is a very conspicuous animal and the most beautifully marked of all the Australian possums. The body has an overall pattern of black and white stripes from the head right down along the bushy tail.

To those experienced with mammals, black and white stripes mean BEWARE. Fortunately, the striped possum does not seem to be able to turn on its scent in quantity, and project it as a skunk does, but those, like David Fleay, who have kept it in captivity, have found its smell very powerful.

Despite the fact that the Striped Possum appears to be common in rain forest, and adjacent more open forest, between Cape York and Townsville, very little is known about its habits beyond a number of observations made on captive specimens by David Fleay, when he was at the Healesville Sanctuary, and A. L. Rand while on one of the Archbold Expeditions for the American Museum of Natural History. From these observations in captivity, the Striped Possum is strictly nocturnal and feeds in a manner very similar to that of that strange primate, the Aye Aye of Madagascar. Both animals have sharp chisel-shaped incisors with which they tear open the bark of trees, and then insert a long thin finger, specialized for the purpose, into the crevice to extract an insect. Although a mixture of foods is taken in captivity, it is probable that the Striped Possum is almost exclusively insectivorous. This possum is an amazingly supple climber, so at ease when head down, that Dr Rand said, 'apparently these animals are so much at home in any position that they do not bother to move to an upright one'. In searching for food they constantly tap their fingers on the bark as though to 'echo-sound' for grubs or termites beneath.

The New Guinea natives claim that the Striped Possums live in hollow trees, while in Queensland rain forest, Robert Grant, a collector for the Australian Museum, found them within Elk Horn and orchid clumps.

The Species   Group 14

STRIPED POSSUM, *Dactylopsila trivirgata* (Plate 20). North-eastern Qld, from the tip of Cape York to near Townsville; rain forest and associated sclerophyll forest.
Recognition: overall pattern of black and white stripes.

20 STRIPED POSSUM, *Dactylopsila trivirgata*   Group 14

Group 15 Pigmy possums   Plate 21

People other than naturalists do not expect possums to be as small as mice, but, even when they are, it is not difficult to distinguish them from mice. Quite apart from their pouches and possum-like upper incisor teeth (which are six in number, unlike the single pair of upper chisels of the mice), all pigmy possums have a prehensile tail which at rest is carried curled around like that of a ringtail. In many of them the tail is also used as a fat store, as it is in the Fat-tailed Dunnart (one of the smallest of the native-cat family) and a number of other marsupials. However, there is no risk of confusing a Fat-tailed Dunnart with a pigmy possum with a swollen prehensile tail because no members of the native-cat family have prehensile tails. Moreover, unlike mice or members of the native-cat family, all pigmy possums are syndactyl (see p. 39). Pigmy possums are insect and nectar-feeders; they often nest in abandoned birds' nests; they may also be found in various situations such as under bark and in hollows. When found they are often in a torpid condition due to their ability to lower their body temperatures and to enter a state similar to hibernation.

Little is known about breeding in the pigmy possums, or even of the litter numbers in the various species. Except in the Mundarda, all species have normally four teats; in the Mundarda, six teats occur and up to six young have been found in the pouch. For many years the Mundarda was thought to be able to delay embryonic development in the manner of some kangaroos and wallabies (see p. 42) but recent research by Mrs Meredith Smith of the University of Adelaide has shown that this is a misunderstanding of the early data.

There are three genera of pigmy possums; the most common is *Cercartetus* which has four species, one of which, the Long-tailed Pigmy Possum, also occurs in New Guinea. It is probably an example of an Australian group which has succeeded in crossing Torres Strait in the opposite direction from that taken by the tree-kangaroos and cuscuses. The other pigmy possums are *Burramys*, the recently discovered living fossil from Mount Hotham in Victoria (see page 14), and the Pigmy Glider or Feathertail *Acrobates* which is discussed elsewhere with the other gliders (page 76) and appears on the one-cent piece.

Until recently, there has been considerable confusion over the identification and distribution of the species of *Cercartetus*; Mr Norman Wakefield of Monash University has now clarified this, however. The distribution given here is the outcome of his work (with the addition of Kangaroo Island to the range of the Little Pigmy Possum; until Mrs I. S. Davis of Brookland Park, Kangaroo Island, S.A., captured a Little Pigmy Possum there in 1964, that species had never been seen in the wild outside Tasmania, although its fossil remains had previously been obtained from the mainland like those of other species nowadays confined to Tasmania such as the Thylacine and the Tasmanian Devil).

The Species   Group 15

Species cannot readily be identified except by means of details of dentition.

MUNDARDA, SOUTH-WESTERN PIGMY POSSUM, *Cercartetus concinnus* (Plate 21). South-west

21　SOUTH-WESTERN PIGMY POSSUM, *Cercartetus concinnus*　　　Group 15

of W.A. (eastwards to Esperance and Coolgardie), south and south-eastern S.A., western
Vic. and south-western N.S.W. (Murray River at Trentham Cliffs); dry sclerophyll forest
with under-storey of shrubs or low trees, or tree and shrub heaths.
Recognition: size of mouse; prehensile tail a little longer than body; colour of upper surface
fawn-brown or grey-brown; under-parts pale, belly fur white; about six to seven inches
total length.

EASTERN PIGMY POSSUM, PIGMY POSSUM, *Cercartetus nanus*. South-eastern S.A., Vic.,
eastern N.S.W., Tas.; wet and dry sclerophyll forest.
Recognition: total length eight to nine inches; belly fur often greyish; otherwise like
C. concinnus.

LITTLE PIGMY POSSUM, TASMANIAN PIGMY POSSUM, *Cercartetus lepidus*. Tas., Kangaroo
Island, S.A.; sclerophyll forest.
Recognition: about six inches in total length; otherwise like *C. concinnus* and *C. nanus*.

LONG-TAILED PIGMY POSSUM, *Cercartetus caudatus*. North-eastern Qld (vicinity of Cairns),
New Guinea; ? rain forest.
Recognition: tail about 1½ times the length of the head and body; total length more than
ten inches.

BURRAMYS, MOUNTAIN PIGMY POSSUM, *Burramys parvus* (Plate 2, p. 15). Vic. (Mount
Hotham). Habitat unknown.
Recognition: as in Plate, rat-sized (larger than all other pigmy possums).

Group 16 The Honey possum   Plate 22

The slender-nosed Honey Possum, or Noolbenger, is, like the pigmy possums, only
the size of a mouse. But they are only distantly related.

The Honey Possum is a nectar and pollen feeder which is highly specialized for this
diet. It has a long tube-like mouth and a tongue which is brushed at its tip like that
of a honey-eating bird. Like these, too, it probably supplements its diet with small
insects which it obtains on blossoms, but accounts which appear in the literature of
Honey Possums catching flies and moths, are probably due to pigmy possums being
misidentified as this species.

Honey Possums normally carry two young in the breeding season, which is probably
in June, July and August. The animal kept by Ella Fry to illustrate this book, revealed
that the young emerge some four months after birth, and are then suckled for a further
two weeks from outside the pouch. At the time of emergence, each young one is nearly
half the length of its mother, and they cease returning to the pouch a few days after
emerging. Our captives lived for seven months before succumbing to a paralytic
deficiency disease, the female having been caught in early July with two very small
hairless young in her pouch. In captivity these possums showed strong preferences
for certain wildflowers such as Menzies' Banksia and bottlebrushes. They would lick
pollen from the stamens as well as drink nectar.

Honey Possums seem to spend quite a lot of time on the ground because they are
most commonly captured after falling into fence-post holes. They appear to have
particular association with the prolifically flowering sandplain flora of the south-west
of Western Australia and, although they are common today, the increasing agri-

22   HONEY POSSUM, *Tarsipes spencerae*   Group 16

cultural use of 'light land' in the south-west threatens them in all parts of their range which is not included within reserves.

The Honey Possum is a zoological enigma. It has no obvious close relatives, and much work yet remains to be done before we understand it, or its ancestry.

The Species Group 16

HONEY POSSUM, NOOLBENGER, *Tarsipes spencerae* (Plate 22). South-west of W.A. (as far north as Murchison River at the coast and Coorow inland; as far east as Salmon Gums and other localities on Esperance–Coolgardie road); tree and shrub heaths.
Recognition: three longitudinal stripes along back; mouse-sized; long pointed nose.

Group 17 The Koala Plate 23

The Koala is one of the most highly specialized of our herbivorous marsupials. It is almost exclusively arboreal although it comes to the ground on occasions to cross from one tree to another. Only a few species of smooth-barked eucalypts are suitable for food trees and, from observing captive stocks, it seems certain that Koalas need a varied diet selected from among these. This may be because some, such as the Manna Gum, produce poisonous prussic acid in new leaves at certain times. In Victoria, the principal food trees are Manna Gum (*E. viminalis*), Messmate (*E. obliqua*), Swamp Gum (*E. ovata*), Peppermint (*E. radiata*), Mahogany (*E. botryoides*). In New South Wales, Forest Red Gum (*E. tereticornis*), Sydney Blue Gum (*E. saligna*) and also the Brush Box (*Tristania conferta*) are eaten. Each animal consumes about two and a half pounds of leaves per day.

Quite apart from its dietary specializations, the Koala is also highly specialized physically for its arboreal existence. Its long arms and curved claws ensure a secure hold, and the vice-like grasp between the first two and last three fingers of the hand is even better developed than that of other possums which employ the same arrangement. At night, in areas where they are numerous, the grunting, pig-like sounds of Koalas can be heard; when they are alarmed or distressed they make a continuous loud wailing sound. Koalas are quite large animals; Victorian males average twenty-three pounds in weight, and females average eighteen pounds.

Of the other Australian marsupials, the Koala seems to be most nearly related to the wombats although the relationship is not close. It appears to be of about the same degree as that by which possums are related to wallabies (i.e. they are different zoological families.)

The Koala is a summer breeder, breeding in November through to February in Victoria, and from September to late January in New South Wales. The young are born about a month after mating occurs, and are then carried in a pouch (which opens backwards) for from five to six months. Towards the end of this period, the young Koala leaves and enters the pouch at will. After the pouch period it is carried on the mother's back for some months more. Koalas mature in the third or fourth year of life, and an animal has been recorded as living for twelve years in captivity. It is not

23   KOALA, *Phascolarctos cinereus*   Group 17

known how often a female Koala reproduces; it is suspected that a female can produce a young each year, but it is not known definitely if she does.

From autopsies performed by Dr A. Bolliger of the University of Sydney on animals found ill in the bush, it is clear that Koalas suffer severely from the dangerous fungal disease cryptococcosis which occurred in a very large percentage of the animals which he examined. This disease is fatal and can affect man; accordingly, Dr Bolliger has advised that close contact with Koalas (and, for example, the almost irresistible temptation to nurse and cuddle them) should not be encouraged. Dr Bolliger showed that the old anecdotal account of Koalas eating soil is true, and he suggests that the source of fungal infection could be in the soil which they eat in this way.

During the late Pleistocene period, Koalas also occurred in the south-west of Western Australia. The reason for their extinction there is not understood, because trees still suitable for food grow in the area. A population of Koalas of mixed origins has been established in captivity in Western Australia, at Yanchep, a National Park to the north of Perth; some of these are the animals figured in Plate 23.

The Species   Group 17

KOALA, NATIVE BEAR, *Phascolarctos cinereus* (Plate 23). South-eastern Qld, eastern N.S.W. Vic., south-eastern S.A.; dry sclerophyll forest, woodland.
Recognition: tail-less; tufted ears; large naked nose.

Group 18 Wombats   Plate 24

There is copious anecdotal information about the natural history of wombats but very little of a precise nature. This is rather strange, because wombats have been very common animals. Moreover they are easy to keep in captivity but so far no one had had success in breeding them; only two cases of successful captive breeding have been recorded in the scientific literature—neither in Australia.

There are two main kinds of wombat. Of these, the best known in eastern Australia, is the Common Wombat, which is the only species in the genus of forest wombats (*Vombatus*). The other genus of plains wombats (*Lasiorhinus*) is only familiar to bushmen, or to travellers across the Nullarbor. It contains three species, the Hairy-nosed Wombat, the Queensland Hairy-nosed Wombat, and the Moonie River Wombat; these inhabit the woodland savannah or grassy plains of the more arid inland.

Wombats are large animals up to about four feet in length, and weighing up to sixty pounds. They have enormous strength and are very efficient diggers, with the result that they are unpopular among people on whose land they occur. The Common Wombat is probably secure as a species because, in part at any rate of its range, it occupies the high mountain forests where the private landholder plays no part in its destruction. Unfortunately, the same is not true of the Hairy-nosed Wombat which appears to be occupying a rapidly-shrinking range and, although it is very common where it does occur, due to its habit of living in densely populated communities, it is very vulnerable. Unless something is done to protect it, its range may shrink beyond

24 COMMON WOMBAT, *Vombatus ursinus*   Group 18

recovery. Already its populations have contracted to the plains of the western side of the Murray River between Swan Reach and Morgan, to the southern border of the Nullarbor Plain to the east of Eucla, and to some small colonies in the Gawler Range, on Eyre Peninsula and Yorke Peninsula in South Australia. Active steps have been taken to protect one colony, that at Blanchetown in South Australia, to the west of the Murray River; there the Natural History Society of South Australia and the owner of Portee Station, Mr B. Powers, have acted together, and by public subscription, have been able to secure 3,000 acres as a special wombat reserve. While this is a highly commendable step in the right direction, it cannot be regarded as adequate and we must continue to work until the remaining populations of the species, and their environments, are fully protected wherever they occur.

The Hairy-nosed Wombats of Blanchetown have been studied in the field, and a few individuals in captivity, by Mr R. T. Wells of the University of Adelaide. His observations show that, during winter and in spring, the wombat is a very selective feeder which grazes on only a few of the herbs available within its range. It seems to be able to do without drinking like the desert-adapted kangaroos and wallabies and, like other marsupials which have been studied, achieves this by means of a combination of factors which are partly physiological and partly behavioural. For example, wombats escape the main heat of the day by living underground in tunnels which are probably fairly humid—they emerge at night to feed and are most active at dusk and at dawn; they only come out of their burrows in broad daylight to sun themselves on warm still days in the cooler part of the year. They also conserve water through being able to concentrate indigestible residues from their food so effectively that they lose a minimum of water in voiding faecal pellets. Mr R. French of the University of Adelaide found that these contain less than fifty per cent of water. By comparison, the faeces of water-deprived camels contain about seventy-five per cent of water. Hairy-nosed Wombats do not seem to be able to sweat either; while this could undoubtedly assist them to conserve water, Wells has pointed out that it must also make them very vulnerable to heat stress. Zoos should be particularly careful not to confine such animals in pens in which they cannot use their behavioural characteristics to escape high temperatures and low humidities.

The studies made by Mr Wells also seem to show that, like the kangaroos of arid places (see p. 50), the Hairy-nosed Wombat is able to subsist on a diet which is low in protein through having the same kind of ability to use micro-organisms to digest otherwise indigestible cellulose; like the Quokka they can possibly even re-use waste urea. In short, these plains wombats are another example of a major division of a marsupial group which has become specially adapted to life under the arid conditions of the Australian inland.

Although many small Australian mammals burrow, wombats are the only large ones which do. Forest Wombats generally dig separate holes with a single entrance situated at the base of a tree or rock. Underground, the holes are branched and are often very deep and long. Finlayson records abandoning one attempt to dig out a Common Wombat when he had reached forty feet from the entrance and six feet below the surface. The Hairy-nosed Wombat digs complex burrow systems comprising a large number of separate burrows which join together to form a warren. Very often the entrances of a number of holes meet together at a kind of crater three or four feet deep, and large enough for a vehicle to fall into. Above ground the entrances of the warren are connected by a network of trails while other trails radiate out to feeding areas and to other warrens. Despite the gregarious appearance which their warrens give them, Hairy-nosed Wombats seem to be essentially solitary in behaviour and to have their own burrows as well as their private feeding areas.

The most complete, and most enthralling, account of wombats and their burrows is by Peter Nicholson who, when he was a schoolboy in 1960 at Timbertop in Victoria, studied Common Wombats which were living in wet sclerophyll forest at between 1,500 and 4,000 feet above sea-level. He did this by entering their burrows. The burrows he investigated were often a network of tunnels; one of them was about sixty feet long. He found that in the burrow there is usually only one wombat, occasionally two, but they are sociable and visit each others' burrows. Wombats are always extending and altering their tunnels. An excavating wombat sits firmly on its rump and hind legs, digs out the earth with its forepaws and thrusts it out to one side; then, as the wombat backs out of the tunnel, the heap of loose dirt is moved out with it by means of the front and back paws. A nesting chamber, lined with vegetation such as broken fronds and bark, is usually situated in half light about six to twelve feet from the entrance. Young Common Wombats learn to tunnel and dig within their mother's burrows. They gradually dig out their own small burrows within the larger ones but about four months after leaving their mother's pouches they leave the maternal burrow altogether.

During his subterranean investigations, Nicholson found that wombats he encountered were generally inquisitive and friendly, only once was he chased out of a burrow by a wombat; it grunted and advanced upon him. One wombat became so accustomed to him that it would follow him out into the open on days when the weather was dull, and it appeared to enjoy being scratched by him.

Common Wombats are probably born in late autumn, and are independent of their mothers by the following summer. They usually have a single young. The Hairy-nosed Wombat also appears to be a summer breeder and, like the Common Wombat, has one young, although twins are known in both kinds.

There is probably only a single species of forest wombat although many naturalists believe that the Tasmanian and island forms should be regarded as separate; but inadequate work has been done to show this. This wombat still occurs in Tasmania and Flinders Island although, at one time, it occurred in King Island, and on Cape Barren Island, in Bass Strait. One of the plains wombats, the Queensland Hairy-nosed Wombat, was formerly plentiful in the Riverina of New South Wales, but none have been reported from there in recent times and Dr Peter Crowcroft, who has most recently examined their status (and upon whose authority we accept them as a distinct species here), says that he could gain no information of any wombat populations between the extant isolated populations at Clermont, in central Queensland, and the former colonies at Deniliquin in New South Wales. Another plains wombat, the Moonie River Wombat, has been recorded as a very restricted population near St George in south-eastern Queensland. The Queensland Museum knows of no record of it elsewhere even in the fossil state, so the status and history of this isolated species is quite unknown.

Forest Wombats have rather short ears (see Plate 24), a coat of coarse thick brown fur and a large naked snout with granular skin on it. By contrast, plains wombats, although obviously wombats from their general shape, possess fine and silky hair and have very much larger pointed ears and a nose covered with soft short hair. Wombats are entirely vegetarian.

The Species

FOREST WOMBATS, *Vombatus*

COMMON WOMBAT, ISLAND WOMBAT, TASMANIAN WOMBAT, *Vombatus ursinus* (Plate 24). South-eastern Qld (south of Stanthorpe), eastern N.S.W., southern Vic., south-eastern S.A., islands of Bass Strait, Tas.; wet and dry sclerophyll, especially rocky country and, in south-eastern S.A. in coastal *Melaleuca* and *Gahnia* complex.
Recognition: coarse hair; bare granular skin on snout; ears short.

PLAINS WOMBATS, *Lasiorhinus*

HAIRY-NOSED WOMBAT, *Lasiorhinus latifrons.* Southern S.A. from the Murray to far south-eastern W.A.; savannah woodland, grasslands, shrub steppe.
Recognition: fine silky hair; nose with fine hair; ears large.

QUEENSLAND HAIRY-NOSED WOMBAT, *Lasiorhinus barnardi.* East-central Qld (vicinity of Clermont), Riverina district of N.S.W. (vicinity of Deniliquin).
Recognition: as for *L. latifrons.*

MOONIE RIVER WOMBAT, GILLESPIE'S WOMBAT, *Lasiorhinus gillespiei.* South-eastern Qld (St George).
Recognition: as for *L. latifrons.*

7

Bandicoots

THE BANDICOOTS comprise the second of the three main branches of marsupials which occur in Australia. Their evolutionary adaptations have fitted them for digging and living on a rather varied diet of invertebrates which they seek in the litter of vegetation, among the roots, and in the earth. A number of species construct burrows but others make nests under heaps of vegetation.

There are two main evolutionary lines of bandicoots. One of these comprises the beautiful silken-furred and long-eared rabbit bandicoots, while the other contains the much more spiny-furred ordinary bandicoots and the Pig-footed Bandicoot.

There is little problem about identifying a bandicoot as a member of the Peramelidae or bandicoot family. First, all bandicoots have the same two little syndactyl toes as the Diprotodonta (see p. 39), and because they live terrestrial lives they have developed elongated hind feet rather like those of the kangaroos and wallabies. But unlike the Diprotodonta none of them has the two procumbent, lower scissor-like incisors of that group. No bandicoot has less than three pairs of little incisors in the lower jaw between the two larger canines. Externally, the least specialized of all the wallabies, the potoroos, look rather like large bandicoots but have diprotodont teeth.

The common name 'bandicoot' is not originally an Australian name; properly, it belongs to a species of large Indian rat, but to give it up on that account would be as ridiculous as to insist on using the word 'Phalanger' for 'Possum' on the grounds that the latter is a corruption of 'possum', the common name of the American Virginian Opossum—a completely different kind of marsupial.

Group 19 Short-nosed bandicoots   Plate 25

Like most bandicoots, the short-nosed bandicoots have harsh, almost spiny, fur which comes out very easily when the animals are handled. The spiny fur consists of specialized outer guard hairs beneath which is a soft under-fur.

All bandicoots have long noses but those of short-nosed bandicoots are shorter than most. It seems likely that bandicoots have such long noses because they feed mostly upon small invertebrates such as insect larvae, worms and spiders which they dig out of the ground; the little conical holes which they make in their avid search for food are a familiar feature of bushland and, in some places, of suburban gardens.

Short-nosed bandicoots make nests like flattened heaps of sticks and debris which are extremely well concealed in vegetation. There is no permanent hole—the animal merely burrows into the nest and then conceals its entrance. Similarly, it emerges by burrowing out; the exit is then covered up. Dr John Kirsch has recently shown that, in south-western Australia at any rate, the Short-nosed Bandicoot will also construct burrows in sandy soil during very hot weather.

There are probably only three species of short-nosed bandicoots although it is difficult to be certain of this without a great deal more study. In southern Australia, the Short-nosed Bandicoot belongs to the species *obesulus*; in the northern and north-eastern areas of Australia it is replaced by the species *macrourus* while the species of the Centre and north-west is *auratus*. The southern and northern species, *obesulus* and *macrourus*, probably once came into contact in the Sydney district. An outlying population of the southern species has been reported in Cape York as well. It is likely that the Barrow Island Bandicoot (which is sometimes regarded as a separate species) is an island form of *auratus*. This is the basis for the simplified classification of short-nosed bandicoots which I give below.

In some parts of Australia short-nosed bandicoots are quite common. For example, in south-western Australia the Short-nosed Bandicoot, locally called the Quenda, is the most common native mammal in the Darling Range behind Perth. Quendas are entirely nocturnal and are seldom seen during the day unless flushed; but many are killed on the roads and their corpses are common on highways leading out of the city. Adults are the size of a rabbit. In the suburbs around Sydney, it is a long-nosed bandicoot, not a short-nosed bandicoot, which is common (see next group).

The most important studies of short-nosed bandicoots to date have been made of a population of Brown Bandicoots in Tasmania by Dr George Heinsohn. He found that they were strongly aggressive and territorial with probably very little overlap of territories of individuals of the same sex. Male territories are larger than female territories, and each male territory probably overlaps with several female territories.

The Brown Bandicoot is secretive and seeks its food close to good cover. At the time of Dr Heinsohn's study, the density of the population was from three to eight individuals per 100 acres. In Tasmania it has two or three litters of up to five young (there are usually two or three) in each breeding season which lasts from May to February. In northern New South Wales and Queensland the Brindled Bandicoot breeds the year round; its gestation period is about two weeks.

The Species   Group 19

BROWN BANDICOOT, QUENDA, SHORT-NOSED BANDICOOT, *Isoodon obesulus* (Plate 25). Eastern N.S.W., southern Vic., Tas., south-eastern S.A., Nuyts Archipelago, south-west of W.A., Recherche Archipelago, Cape York Peninsula; sclerophyll forest, woodland, heaths, wherever there is good ground cover.
Recognition: as in Plate; ear rounded; back dark greyish or yellowish brown, darkly grizzled; upper surfaces of fore feet greyish or brownish.

BRINDLED BANDICOOT, LONG-TAILED SHORT-NOSED BANDICOOT, *Isoodon macrourus*. North-eastern N.S.W., eastern Qld, Arnhem Land, north Kimberley; dense ground cover in sclerophyll forest or woodland but not rain forest.
Recognition: ears rounded; N.S.W. and Qld animals upper-surface brown brindled black, Arnhem Land and Kimberley animals paler; fore feet white, rest of leg brownish; hind feet brownish or greyish-white.

GOLDEN BANDICOOT, WINTARRO, *Isoodon auratus* (including *I. barrowensis*). Central Australia from north-western Australia through N.T. and inland S.A.
Recognition: ears rounded; golden-brown back.

25 BROWN BANDICOOT OR QUENDA, *Isoodon obesulus*   Group 19

Group 20   Long-nosed bandicoots   Plate 26

There are two genera of bandicoots in Australia which have particularly long noses. One of these (*Echymipera*) contains only a single species, the Spiny Bandicoot; but in New Guinea this genus is represented by three species, one of which is identical with that occurring in Australia. The Australian population of the Spiny Bandicoot is only known from a single individual which was collected by Dr P. J. Darlington of the Harvard University Expedition, in 1932, in the dense rain forests of the Upper Nesbitt River which flows down the eastern slopes of the McIlwraith Range in Cape York. Despite intensive searches by subsequent expeditions, no more have been found.

The fur of the Spiny Bandicoot is particularly coarse and spiny; the bristles are flattened and are channelled on each side; and even the under-fur is coarse.

The other genus of long-nosed bandicoots (*Perameles*) contains a number of the much commoner Australian species. While their fur is not as harsh as that of the Spiny Bandicoot, it is still quite spiny like that of the short-nosed bandicoots; but by comparison with short-nosed bandicoots they are rather long-eared and are more delicately built and longer of limb. All but one of the species, the Common Long-nosed Bandicoot of eastern Australia, has a clearly marked pattern of bars across the hind quarters; juvenile specimens of the Common Long-nosed Bandicoot showing a faint barred pattern on the rump are common.

The long-nosed bandicoots are nocturnal and carnivorous, feeding principally on small invertebrates such as insect larvae, earth worms, and spiders; they will also take berries when these are available.

Much of our knowledge of the breeding of long-nosed bandicoots comes from studies of the Common Long-nosed Bandicoot by Dr Gordon Lyne and by Dr Eleanor Stodart, and of Gunn's Bandicoot in Tasmania by Dr George Heinsohn.

In the Sydney populations of the Common Long-nosed Bandicoot, breeding takes place all the year round and young are born after a gestation period of about two weeks. There may be litters of up to five, but two or three are normal. The young leave the pouch permanently at an age of nine to ten weeks. Young, half-grown, females start breeding at about four months; males become sexually mature at five months. They are nocturnal and extremely solitary; males being aggressive when they come into contact, and females generally avoiding each other (and males when they are not in season). When a female is attractive to a male she will be followed until mating takes place—as Dr Stodart has put it, '*Perameles nasuta* is a solitary animal and interaction is almost restricted to the minimum necessary for successful reproduction . . .'.

George Heinsohn found very much the same pattern of behaviour in Gunn's Bandicoot. He found that females have average ranges of about eight acres, but these are quite variable and there may be considerable overlap in female ranges; males have much wider ranges of between forty-six and ninety-eight acres, so that, presumably, one male includes a number of females in his territory. Like the Common Long-nosed Bandicoot, Gunn's Bandicoot behaves in a solitary manner unless the female is in season.

In Gunn's Bandicoot, the young leave the pouch at seven weeks, being weaned at between eight and nine weeks. The female becomes sexually mature at about three

LONG-NOSED BANDICOOT, *Perameles nasuta*   Group 20

months, and, during the breeding season from May to February, three to four litters are produced by the female. Litters of up to four have been recorded, but two to three are normal, and females born early in the season grow rapidly enough to bear one or two litters during the same season. It is not known what keeps populations of bandicoots in check but they appear to have large numbers of natural predators. Density of the population studied by Dr Heinsohn seemed rather variable for reasons that are not known; over a study period of thirteen months numbers went down from fifty-nine to eighteen per 100 acres.

The Species   Group 20

LONG-NOSED BANDICOOT, *Perameles nasuta* (Plate 26). Eastern Qld, eastern N.S.W., eastern Vic.; rain forest, wet and dry sclerophyll forest, sclerophyll woodland.
Recognition: without bars; ears with pointed tip; fur harsh, grey or brownish grey, hind feet white, fore legs white to elbow.

GUNN'S BANDICOOT, TASMANIAN BARRED BANDICOOT, STRIPED BANDICOOT, *Perameles gunnii*. Southern Vic., Tas.; woodland and open country with good ground cover.
Recognition: tail white below and above except for thin greyish or brownish patch on upper surface for about an inch close to rump; pronounced bars across hind quarters.

MARL, LITTLE MARL, BARRED BANDICOOT, *Perameles bougainville* (following Wakefield, 1966, this includes *fasciata*). Arid southern Australia—from islands in Shark Bay through W.A. south of the Tropic (in the South West probably in sandplain only), southern S.A., north-western Vic., western N.S.W. to Liverpool Plains; heaths, dune vegetation.
Recognition: tail dark above; bars across hind quarters not strongly marked.

DESERT BANDICOOT, *Perameles eremiana*. Northern S.A., southern N.T., Great Victoria Desert, W.A.; spinifex grasslands.
Recognition: general colour above dull orange, fur soft with dark spiny guard hairs; under-surface, hands and feet, white; ears long, narrow and pointed reaching past eye when laid forward.

SPINY BANDICOOT, *Echymipera rufescens*. McIlwraith Range, north-eastern Cape York Peninsula, Qld; rain forest.
Recognition: eight small but rather broad incisor teeth between upper canines (other bandicoots have ten); soles of hind feet bare of fur to the heel (other long-nosed bandicoots have rather hairy heels, but this may be worn in old specimens).

Group 21 The Pig-footed bandicoot   Plate 27

To my knowledge, no modern biologist has ever seen a living Pig-footed Bandicoot. In fact, the most reliable recent record of it seems to be information from elderly Pitjanjarra men in the Musgrave Ranges, which was given to Mr H. H. Finlayson; they have not seen it since about 1926.

This bandicoot is a very graceful little mammal, somewhat smaller than a rabbit. It has a slightly crested tail, elongate, delicate ears, and slender legs rather like a miniature deer. Its name 'Pig-footed' comes from the resemblance between its fore feet and the cloven trotters of a pig. In the fore foot of the Pig-footed Bandicoot, only the second and third toes are well developed (unlike a normal bandicoot the fourth toe is

7 PIG-FOOTED BANDICOOT, *Chaeropus ecaudatus*   Group 21

minute, while the first and fifth are absent altogether); the hind foot mostly consists of the greatly enlarged fourth toe, while the first toe is absent and the two syndactyl toes (see p. 39) are very small. The fur is grey-brown, coarse and straight, but not spiny. The low crest on the tail is black.

There is little information on the natural history of the Pig-footed Bandicoot and much is conflicting, although some of the apparent conflict (like the account in Sturt's exploration of Central Australia) is probably due to misidentification.

The pouch opens backwards and contains eight teats, although reliable accounts state that no more than two young are carried. As far as can be construed, the breeding season seems to be in winter.

Gerard Krefft, who collected them on the Murray in 1857, regarded them as being principally vegetarian, and said that they did not attack mice like other bandicoots, or eat meat.

Krefft's, P. M. Byrne's and John Gilbert's observations suggest that the Pig-footed Bandicoot is nocturnal, spending the day in a low, covered nest of dried leaves, pieces of stick, dried grass, etc., with a lining of soft fibrous grasses. It burrows its way out of the nest, leaving no opening, and when startled from its nest in daylight, will take refuge in a hollow log. John McKenzie, who collected a Pig-footed Bandicoot for Dr E. C. Stirling in 1907, in a letter describes the nest in Central Australia as a hole filled in with drygrass under which the bandicoot crawls. The hole is a short burrow about a foot long, about eight inches across and eight to ten inches deep and, apart from its filling of dried grass, is left open.

The Species   Group 21

PIG-FOOTED BANDICOOT, *Chaeropus ecaudatus* (Plate 27). South-western N.S.W., Vic. (near junction of Murray and Murrumbidgee Rivers), N.T. (south of 21° South), northern S.A., W.A. (near Northam); sclerophyll woodland, mallee, heath, and grassland.
Recognition: brownish above, pale below; slender limbs with two functional toes only on fore feet; long tail with black crest; ears long and narrow.

Group 22 Rabbit-eared bandicoots
bilbies or dalgytes

Plate 28

One has only to see a living Rabbit-eared Bandicoot to be convinced that it is one of the most beautiful and graceful of our native mammals. Its hair is long, soft and silky, and its tail, so strikingly marked in black and white, is carried like a banner. Its progression is a graceful flowing canter.

Unlike the other bandicoots, the Rabbit-eared Bandicoot always lives in a burrow and is a most accomplished and powerful digger. Each fore paw, armed with powerful claws, scratches backwards independently and then the earth loosened in this way is thrown back violently by the hind feet working together. By this means an incredible amount of earth can be displaced in a very short time. Their burrows are from three to six feet deep and descend spirally. By living in these deep burrows they escape the heat stress of the central deserts in which much of the population lives. As far as is

28   BILBY OR DALGYTE, *Macrotis lagotis*

Group 22

known the burrows have only a single entrance, which is marked by a mound of excavated dirt, although the entrances themselves are often concealed by spinifex or other low vegetation. D. R. Smyth and C. M. Philpott in 1966 studied the species near the Warburton Range, Western Australia, on behalf of the Royal Zoological Society of South Australia, and found that burrows were most common in shrub grasslands comprising mulga, spinifex and tussock-grasses. Burrows containing Rabbit-eared Bandicoots which were dug out were found to be blocked at intervals by soft earth and no burrow contained more than a single bandicoot. In the area of a population investigated by Smyth and Philpott they located fifty-eight burrows and none was more than 550 feet from any other. At the most only twenty-seven per cent of these were used on any one night. It seems likely that the Rabbit-eared Bandicoot is a solitary animal, like other bandicoots, and also that each has many burrows in its range which it visits.

Rabbit-eared Bandicoots are carnivorous animals, accepting both meat and insects. It is reported that they extract beetle larvae from the roots of acacias, and in the Northern Territory, Dr A. E. Newsome found that the principal food in the wild was termites. They are very variable in size; some are larger than a cat, while other individuals are very much smaller. They are quiet captives, but scratch fiercely. They sleep sitting back upon their tails with their heads tucked down between their fore paws and ears folded forward along the face.

The breeding season appears to be from March to May; the pouch, like that of all bandicoots, is backward-opening and the litter size is from one to three.

The distribution and classification of Rabbit-eared Bandicoots is not clearly understood. The distribution seems always to have been very patchy, but the animals were formerly extremely common where they did occur. Thus, for example, dalgytes were very familiar animals of the wheat belt of Western Australia where now they are completely unknown. They were last reported in Victoria in 1866 and in New South Wales in 1912. Populations still occur in remote and rather arid places such as Dampier Land, the Pilbara and in desert areas of central Australia including south-western Queensland, parts of the Northern Territory and the western desert.

There are generally thought to be only two species, the Common Rabbit Bandicoot and the small Central Australian Yallara or Lesser Rabbit Bandicoot. It is possible, however, that the strikingly-marked Black-footed Rabbit Bandicoot from Ooldea in South Australia may be a distinct species and not merely a local variety.

The Species   Group 22

COMMON RABBIT BANDICOOT, BILBY, DALGYTE, *Macrotis lagotis* (Plate 28). W.A. (between the South West and Dampier Land), S.A., N.T. (north to about 16°S), N.S.W. (west of the Divide, south-west to the Murray and Darling), south-western Qld; woodland and plains including savannah woodland, open tree savannah and shrub grasslands.
Recognition: rabbit-like ears, tail black and white with prominent white crest.

YALLARA, LESSER RABBIT-EARED BANDICOOT, *Macrotis leucura*. Central Australia in the vicinity of the Diamantina, Barcoo, and the Lake Eyre Basin; burrows only in sandhills, never on flats; entrance blocked from within by occupant.
Recognition: under-surface of tail white along its entire length.

8

Marsupial carnivores:

NATIVE-CATS AND THEIR RELATIVES

THE NATIVE-CATS and their allies make up together the Australian members of the marsupial carnivores (the others occur in North and South America); they are the third main group of Australian marsupials. The few fossil marsupials known from outside Australia from the beginning of the Age of Mammals indicate that, of all the main Australian groups, these are probably most like the original marsupial immigrants into the continent. As a result, zoologists regard them as the stem forms from which the other kinds of Australian marsupials radiated.

Unfortunately, almost nothing is known of their fossil history within Australia except during the period of the last half million years or so. These fossils add nothing to our knowledge of the radiation of marsupial carnivores that we do not already know from the living forms.

As would be expected from a group of such antiquity, the species are very diverse in their ways of life. The largest and most distinct is the Thylacine, a wolf-like marsupial which nowadays is restricted to Tasmania but which, in the early days of Aboriginal Australia, formerly occurred all over the continent and even lived in the Highlands of New Guinea. The spotted native-cats and the fierce-looking Tasmanian Devil are next in size, and then there is a whole host of smaller marsupials which range in size from less than that of a very small mouse upwards. Some are arboreal while others are even burrowers. All are carnivorous predators, the small ones on insects, the larger ones on a mixture of insects and small vertebrates (like lizards), and the very largest probably prefer mammals and birds.

All marsupial carnivores have three pairs of lower incisors; they all have five toes on the front feet and never less than four on the hind feet, and the second and third toes of the hind feet are separately developed. Thus, they are neither diprotodont nor syndactyl (see p. 39). As a result there is no possibility of the larger ones being confused with other marsupials; but they are often confused with other kinds of mammals, and, in particular, with introduced non-marsupial carnivores. This is because the pouches of most marsupial carnivores are not pocket-like in the popular pattern of the marsupial pouch; in fact, the pouch is usually no more than a saucer-like depression which contains the teats and the small young. Out of the breeding season, the pouch has scarcely any rim to it and it is often filled with rather long gingery hair and is difficult to see. Adult males are easier to identify as marsupials; in all marsupials the scrotum is pendulous and hangs down on a rather long stalk below the belly. The penis is behind it.

The small ones, and in particular the antechinuses and dunnarts, are rather apt to be misidentified as mice. However, once you have seen one of these you will never mistake the alert foxy face of the little marsupial carnivore for the blunter face of a rodent; besides which, all rats and mice have only one pair of chisel-shaped incisors above and below. Moreover, these are often yellow. No marsupial has yellow or orange teeth.

The Marsupial-mole is put into this chapter for convenience. But in honesty it must be pointed out that this is guesswork; the Marsupial-mole is so modified in response to its highly specialized existence that all traces of features which would indicate its ancestry have eluded researchers to the present time. The Marsupial-mole has many incisors, therefore it may be a bandicoot, or it may belong with the native-cats; but its toes are so modified as shovels that we cannot be certain whether it is syndactyl or not. It could even belong in a group on its own.

Group 23 Native-cats   Plate 29

There are four species of native-cat, or *Dasyurus*, in Australia. They are all lean, short-legged, active, medium-sized, predatory mammals and each of the principal Australian faunas (see pp. 1-4) contains at least one of the species. These are the Quoll, the Chuditch, the Tiger Cat, and the Satanellus (or Little Northern Native-cat).

The Australian native-cats are spotted with numerous white blotches, usually on an olive-brown background; this background colour is rather variable having some-times more yellow in it. In the Quoll there are two colour phases, one of which is black and less common. The black variety is always spotted like the brown one.

The native-cats are held by poultry-keepers to be vicious killers. There is no doubt that they raid poultry yards when they live close to man, but, in such situations, they probably also do a great deal of good by taking mice and quantities of pest insects as well. In a recent survey, Mr R. H. Green of Launceston has found that most of the Quolls he collected in agricultural land contained Corbie grubs, a common pasture pest. In fact they are opportunistic feeders which take flesh when opportunity offers, but also take vegetable food, insects and earthworms.

Not a great deal is known about the breeding of native-cats. All appear to be winter breeders, producing young between May and August. The Quoll raises litters of between six and eight, following a gestation period of between eight and fourteen days; the young are weaned at four and a half months and become mature by the time they are a year old. The Tiger Cat has about five young after a gestation period of twenty-one days. As in the Quoll, the young are weaned at four and a half months, and become mature at the end of a year. Less is known about the Satanellus, or Little Northern Native-cat. It has up to eight young which are weaned at some time over three months. They, too, probably become mature at the end of a year. The Chuditch has up to six young.

The young of most species of the native-cat family, once they leave the pouch, still cling and clamber over their mother for some weeks. During this period they suckle with decreasing frequency. Both adults and young are very active climbers.

One of the few known examples of superfetation (or the over-abundant production of young) in Australian marsupials is provided by the Quoll, and, from the numbers

29   WESTERN NATIVE-CAT, *Dasyurus geoffroii*   Group 23

of embryos observed in the uterus, it seems to be the usual condition. Dr Hill and Dr O'Donohue recorded cases of two females giving birth to eighteen and ten young; the pouch contains only six teats—those that do not attach to them die. Superfetation appears to be common in the American opossum *Didelphis*.

The Species   Group 23

TIGER CAT, *Dasyurus maculatus.* Eastern Qld, eastern N.S.W., eastern and southern Vic., south-eastern S.A., Tas.; sclerophyll forest, rain forest.
Recognition: brown body and tail, both spotted with white; five toes on hind feet, ridges on foot pads; adults cat-sized; other *Dasyurus* smaller.

QUOLL, EASTERN NATIVE-CAT, *Dasyurus viverrinus.* Eastern N.S.W., eastern and southern Vic., south-eastern S.A., King Island, Tas.; dry sclerophyll forest.
Recognition: no spots on tail but tip may be white; four toes on hind foot, foot pads granular; may be olive-brown or black spotted with white.

CHUDITCH, WESTERN NATIVE-CAT, *Dasyurus geoffroii* (Plate 29). W.A. from the south-west to south of the Fitzroy River, S.A., N.T. and Qld south of about 20°, N.S.W. west of the Divide, north-western Vic. sclerophyll forest, woodland (presumably also desert associations as well).
Recognition: no spots on tail; usually five toes on hind feet, foot pads granular.

SATANELLUS, LITTLE NORTHERN NATIVE-CAT, *Dasyurus hallucatus.* W.A. (Kimberley and north-west as far south as Roebourne), N.T. (Darwin area and Arnhem Land, Groote Eylandt), northern and north-eastern Qld (as far south as Gympie); forest and woodlands, rocky outcrops.
Recognition: no spots on tail; tail often sparsely haired; five toes on hind feet, ridges on foot pads.

Group 24 The Tasmanian Devil   Plate 30

The black and white, or sometimes all black, Tasmanian Devil cannot be mistaken for any other marsupial. It is about the size of a Corgi and has a large head and fore quarters, and powerful jaws. Large males weigh up to twenty-six pounds. Individuals sound and look incredibly ferocious.

Tasmanian Devils are nowadays confined entirely to Tasmania, although fairly recent fossil deposits on the mainland contain their bones. The only living specimen to be collected on the mainland was taken in 1912 at Tooborac, sixty miles from Melbourne. It has generally been supposed to be an escape from captivity—but this has not been substantiated.

Devils are slow and clumsy in their movements, and their habit of continually nosing the ground probably indicates that they possess a well-developed sense of smell and hunt by it.

The breeding season of the Devil is the end of May or June and up to four young are born; the pouch has four teats. The young Devils leave the pouch fifteen weeks after birth and like native-cats cling to their mother. They are weaned at five months and probably breed at the end of their second year.

Tasmanian Devils make a fearsome noise when they are fighting or threatening an

30   TASMANIAN DEVIL, *Sarcophilus harrisii*   Group 24

intruder, but those who have had experience of domesticated Devils, reared in captivity, have called them 'delightful'. The scientific name *Sarcophilus* means flesh-lover—an appropriate name for a most efficient scavenger of carrion. Devils eat almost everything of a carcass, bones and all, except for certain small bits such as the parts of jaw bones which bear teeth. Devils are also known to eat fish, crustacea and reptiles. They make dens under stones, in cavities, and in old stumps; they are competent climbers.

They have become extremely common animals in Tasmania, after a period of rarity which followed the spread of settlement. In recent years their predation upon domestic stock has made it necessary for control operations to be authorized. As the result of a study by Mr R. H. Green of the Queen Victoria Museum, Launceston, made concurrently with one such operation, Mr Green has expressed the opinion that it is doubtful whether Devils will kill grown sheep in a normal healthy condition, but they quickly recognize the sick or dying animal as an easy kill, and that weak or motherless lambs provide easy prey. Tethered or closely penned sheep are always liable to nocturnal attack and he gave an instance of a sheep dog pup, belonging to an employee of the Icena estate in north-eastern Tasmania, which was killed and completely eaten while on its chain.

The Devil is a close relative of the spotted native-cats.

The Species   Group 24

TASMANIAN DEVIL, *Sarcophilus harrisii* (Plate 30). Tas., possibly southern Vic.; sclerophyll forest with rocky places.
Recognition: size of Corgi; black with white chest or rump markings, or black only.

Group 25 Tuans or wambengers   Plate 31

Tuans or wambengers are brushy-tailed carnivorous marsupials about the size of a rat, but are quite unlike rats in appearance. The name 'marsupial-rat' which is sometimes used for them has not even superficial similarity to commend it. They are sometimes called phascogales, a name which is an anglicized version of the scientific name *Phascogale*.

The Tuan, or Common Wambenger as it is called in Western Australia, is a grey animal with a striking black brush which is often carried fluffed out like a bottle-brush. It is reputedly very savage but individuals which I have kept in captivity have been very gentle and easy to handle. Both David Fleay and Norman Wakefield have remarked on their somewhat timid behaviour when faced by Sugar Gliders. The Tuan is almost exclusively arboreal. It climbs with great agility even descending trees head first, or remaining stationary head downwards with the body pressed flat against the trunk. Sometimes, when the animal is not excited, the tail is carried with the hairs close together as is shown in the lower posture in the Plate. This is always the condition of the tail in dead specimens. An adult animal measures about eighteen inches from nose-tip to tail-tip.

31 <small>TUAN OR WAMBENGER,</small> *Phascogale tapoatafa*   Group 25

The Red-tailed Wambenger is like a smaller version of the Tuan, but is brown with a red base to its tail. Like the Tuan it is a most agile climber and acrobat; one kept captive by Mr D. G. Bathgate, at the Western Australian Museum, would do a continuous succession of back-flips from the surface of a table.

The Tuan seems to be a winter breeder in southern Australia although very small pouch-young have been found as late as 15 November by John Calaby in north-eastern New South Wales. From three to six young have been recorded; lactation lasts for four months and the young appear to be mature at seven months.

It is not known whether there is a restricted breeding season in the Red-tailed Wambenger. The wambengers are arboreal and appear usually to make their nests in hollows; they will feed on nectar as well as insects and flesh.

The Species   Group 25

TUAN, COMMON WAMBENGER, BRUSH-TAILED PHASCOGALE, *Phascogale tapoatafa* (Plate 31). W.A. (the South West, and north-east of Kalgoorlie eastwards to the border, Kimberley), south-east of S.A., southern Vic., eastern N.S.W., eastern Qld (as far north as Mackay), N.T. (northern including Arnhem Land); rain forest, sclerophyll forest, woodland, savannah woodland.
Recognition: blue-grey with black brush.

RED-TAILED WAMBENGER, *Phascogale calura*. Inland parts of the South West of W.A., N.T. (Macdonnell Range), Murray districts of S.A., north-western Vic. and south-western N.S.W.; sclerophyll woodland.
Recognition: reddish-brown with that part of tail between brush and body bright rufous.

Group 26 The Kowari   Plate 32

The Kowari is a brushy-tailed carnivorous marsupial superficially like the Tuan in appearance. It is a light yellow-brown in colour with a striking black brush on its tail. It is confined to the eastern part of the Centre, seems to be entirely terrestrial and lives in holes in the ground. As far as its known characteristics have been studied it seems to be a relative of the spotted marsupial-cats rather than the Tuan.

Some information is available about breeding in the Kowari from a number of specimens which were kept in captivity by the late Mr George Mack, formerly the Director of the Queensland Museum. They bred from May to October, usually having five young, although one female reared six. The gestation period is thirty-two days and males had to be removed before the young were born. Specimens were mature at the end of twelve months.

The Species   Group 26

KOWARI, BYRNE'S POUCHED MOUSE, *Dasyuroides byrnei* (Plate 32). Central Australia (area around junction of N.T., S.A. and Qld); desert associations and grasslands.
Recognition: yellowish or reddish-fawn with contrasting black brush on tail; four toes only on hind foot.

32   KOWARI, *Dasyuroides byrnei* Group 26

Group 27 The Mulgara

Plate 33

The Mulgara, which was called Canning's Little Dog by the men of the expedition led by Canning to survey and open up the great stock route from Wiluna to Halls Creek across the Great Sandy Desert in Western Australia, is a rat-sized carnivorous marsupial which, like the Kowari, is terrestrial and lives in holes in the ground. Unlike the Wambengers and the Kowari, however, it does not have a long brushy tail. Its tail is short and crested.

Recent studies on the Mulgara by Dr K. Schmidt-Nielsen and Dr A. E. Newsome have revealed that, despite the fact that it is a carnivorous mammal which lives in the most arid regions of Australia, its physiological adaptations are such that it is able to subsist without drinking water, or even eating succulent plants, because it is able to extract sufficient water from even a diet of lean meat, or mice. A Mulgara will eat from twenty to twenty-five per cent of its own body weight in meat each day; this diet contains adequate water but, in addition, being high in protein, results in the production of a great deal of urea as well which has to be excreted. In most mammals the major source of water loss is through the water required and lost in urine formation, but the kidneys of the Mulgara are so efficient that they are able to excrete the urea in concentrated form, using very little water. Mulgaras also avoid exposure to heat by remaining in their burrows during the heat of the day.

The Mulgara seems to be a winter breeder and six or seven young are produced after a gestation period of approximately thirty days. Lactation lasts for three and a half months.

Little is known of the relationships of the Mulgara. The characters of its skull and teeth ally it with the Kowari, but its general appearance is more that of a rather unusual *Antechinus* (see the next Group).

The Species

MULGARA, CANNING'S LITTLE DOG, CREST-TAILED MARSUPIAL MOUSE, *Dasycercus cristicauda* (Plate 33). Arid Australia from the Pilbara in W.A. through to south-western Qld, sand-ridge or stony desert, spinifex (*Triodia*) grassland.
Recognition: thick-set body, broad head; reddish or sandy brown above with contrasting crest of shining black hair along upper part of end of tail.

Group 28 Antechinuses

Plate 34

These are often called marsupial-mice or broad-footed marsupial-mice in books, but naturalists, and professional scientists, are more and more coming to call them by the anglicized version of the generic name *Antechinus*. Antechinus (pronounced Anti-kynus) is derived from the Greek words *Anti-* resembling, and *Echinos-* a sea urchin or a hedgehog; it refers to the bristly look of the fur.

3 MULGARA, *Dasycercus cristicauda*

Smaller than the wambengers, the antechinuses are secretive and are seldom seen by people unless they are caught and brought into houses by domestic cats. Unlike the wambengers, they have no brushes on their tails. There are five principal sorts of antechinus in Australia; two of these have only one species each, while the other three have several.

The most common of the sorts of antechinus is the section of the group related to the common Yellow-footed Antechinus, or Mardo. They are active climbers, around both rocks and trees, in rain forest and wet and dry sclerophyll forest. They are light to dark-brownish in overall colour; their hair appears slightly bristly and, in their climbing posture, their hind feet tend to be rotated outwards so that they stick out sideways. The Yellow-footed Antechinus and the Brown Antechinus are very common in eastern New South Wales.

The least-known and most misidentified sort of antechinus is the species called Godman's Antechinus from the mountain rain forests of Ravenshoe near Cairns in Queensland. Most of the material in collections which has been described as belonging to this species belongs to the Yellow-footed section. True *Antechinus godmani* is probably most closely related to some of the interesting species which occur in New Guinea.

Two of the Western Australian species, the Dibbler (see Plate 4) and the Little Red Antechinus, form the third section. They appear to be adapted for life under arid and semi-arid conditions. They have rather short hairy tails which taper rapidly from a stout root to a fine, well-haired tip.

The Dusky (or Swainson's) Antechinus and the Swamp Antechinus comprise the fourth section. The Dusky Antechinus is confined to the wet forests of eastern and south-eastern Australia where it lives on the ground, burrowing under litter. In life, it has the appearance of being a very much 'flattened' animal.

The final section of antechinus contains a species adapted for life among rocky hills in the desert. This is the Red-eared Antechinus which has greyish-fawn, or reddish, fur with slate grey under-fur. It is most easily recognized by its tail which is used as a fat store and is usually swollen and carrot-shaped; further, as its name suggests, living specimens, and fresh skins, have a distinct patch of reddish hair behind each ear. Several names have been given for antechinuses of this section; among these is *Antechinus bilarni* which was named by Dr David Johnson of the U.S. National Museum after Bill Harney who guided the American/Australian Arnhem Land Expedition. This is a long-tailed member of the section and was thought to be a distinct species, but recently Harry Butler has collected specimens in north-western Australia which indicate that, in the northern part of its range, the Red-eared Antechinus has a longer tail than it has in the south. I believe that they should all be regarded as a single species.

Important biological work which throws light on the natural history of *Antechinus* has been done in recent years by B. J. Marlow of the Australian Museum, R. M. Warneke of the Fisheries and Wildlife Department of Victoria, and Dr Patricia Woolley of La Trobe University. All have succeeded in breeding different species of antechinuses in captivity; all have shown that the breeding season is restricted to a short period each year.

In the Brown Antechinus mating occurs in August and, usually, either six or seven young are born in September following a gestation period of thirty to thirty-three

4   YELLOW-FOOTED ANTECHINUS, *Antechinus flavipes*

days; they reach sexual maturity in the year following their birth. In the Yellow-footed Antechinus, mating takes place (in the laboratory, at any rate) in the winter months, between July and September. Wild-caught females normally have ten or twelve young—the full number for which their teats can provide attachment and nourishment. Gestation is from twenty-six to thirty-five days.

Mating in the Brown Antechinus and the Yellow-footed Antechinus is a violent procedure. The powerful males seize the females by the scruff of the neck and manoeuvre them into position. Basil Marlow has described how non-receptive females, unable to escape in captivity from the savage and violent males, are badly lacerated about the head and neck in attempted courtship; they even die from their injuries. Copulation is very prolonged, lasting more than five hours and sometimes up to twelve hours. Females copulate more than once during a mating period which extends over several days; the maximum number of copulations by a single female, recorded by Dr Woolley, was ten, and the maximum duration of the mating period eleven days.

After the breeding season, females recover their condition and may reproduce in a second season. Males, on the other hand, become lethargic after mating, and if they survive in the wild to a second year, they do not recover their vigour.

The Species   Group 28

SECTION 1

YELLOW-FOOTED ANTECHINUS, YELLOW-FOOTED MARSUPIAL-MOUSE, MARDO, *Antechinus flavipes* (Plate 34). Eastern Australia from Cape York to Vic., south-eastern S.A., south-western W.A.; rain forest, sclerophyll forest and woodland.
Recognition: head and fore parts dark grizzled grey, lower back and rump browner or more reddish, flanks reddish-brown; upper-surface of feet off-white to yellowish-brown; often with yellowish-brown eye ring; tail moderately hairy.

BROWN ANTECHINUS, STUART'S ANTECHINUS, MACLEAY'S MARSUPIAL-MOUSE, *Antechinus stuartii*. Eastern Qld (south from Cairns district), eastern N.S.W., Vic.; rain forest, wet and dry sclerophyll forest, woodland, sandstone cave country.
Recognition: uniform drab brown above; no conspicuous features; feet more or less like body, never very pale as in *A. flavipes*; tail moderately hairy.

FAWN ANTECHINUS, FAWN MARSUPIAL-MOUSE, *Antechinus bellus*. N.T. (northern including Arnhem Land); tropical savannah woodland.
Recognition: uniform fawn-grey, or grey above, very pale grey below; upper-surface of feet pale; tail moderately hairy, mostly pale with greyish-brown stripe along its upper-surface.

SECTION 2

GODMAN'S ANTECHINUS, *Antechinus godmani*. North-eastern Qld (vicinity of Ravenshoe, Cairns region); mountain rain forest.

Recognition: tail sparsely haired above, longer haired below to form crest along under-surface.

SECTION 3

DIBBLER, *Antechinus apicalis* (Plate 4, p. 21). Inland periphery of south-west of W.A.; sclerophyll woodland, to shrub and tree heaths.

Recognition: freckled whitish on brown; pronounced white ring around eyes; tail with long hairs, especially close to rump, giving appearance of thick base.

LITTLE RED ANTECHINUS, *Antechinus rosamondae.* Pilbara of north-western W.A. from Port Hedland district to Onslow; spinifex grassland.
Recognition: glossy golden red; tail long-haired, especially close to rump, giving appearance of thick base.

SECTION 4

DUSKY ANTECHINUS, SWAINSON'S ANTECHINUS, DUSKY MARSUPIAL-MOUSE, *Antechinus swainsonii.* Eastern N.S.W., eastern and south-western Vic., coastal south-eastern S.A., Tas.; rain forest or wet sclerophyll forest, alpine woodland, alpine heaths—in deep litter in dense vegetation.
Recognition: head and body very dark brown above; fore claws very long; ears very short.

SWAMP ANTECHINUS, LITTLE TASMANIAN MARSUPIAL-MOUSE, *Antechinus minimus.* Coastal south-eastern S.A., coastal southern Vic., Tas., islands of Bass Strait; tussock grassland, coastal complex.
Recognition: head and body, grey-brown above; fore claws very long.

SECTION 5

RED-EARED ANTECHINUS, FAT-TAILED MARSUPIAL-MOUSE, *Antechinus macdonnellensis.* Pilbara, central desert areas, and Kimberley of W.A., N.T. including Arnhem Land; rocky hills and breakaways.
Recognition: tail usually fat; chestnut patches behind ears; head flattish and broad.

Group 29 Pigmy antechinuses

Plate 35

Among the small carnivorous marsupials there are four species of very small animals indeed. They seem to be pigmy derivatives of *Antechinus.* Three of them are tiny animals—the smallest marsupials known, while the other is not very much larger, being much smaller than a common mouse. The three tiny species have rather flat heads and are often called flat-skulled marsupial-mice, and are commonly put together in one genus, *Planigale.* The fourth species (*maculatus*) is conventionally placed with other *Antechinus.* The relationships of these minute marsupials is largely a matter of conjecture at present. The flat-headed nature of the group is not absolute and some other larger *Antechinus* are flat-headed too. In this book, because they are all tiny, they are grouped together and I only follow the usual convention in the generic names with considerable reservations. In addition, careful examination of the skulls of specimens in collections reveals that there are probably more species than the four usually recognized. At present, some specimens can only be identified as being 'close' to one or other of the described species.

Very little is known about the natural history of the pigmy antechinuses. David Fleay has kept Ingram's Planigales in captivity having received an adult female with seven newly-furred young in February 1960. She made a small, neat, saucer-shaped nest of dried grass between concave shells of wood and bark and ate enormous quantities of insect foods. She consumed more than her body weight in grasshoppers daily, and would also eat small dead birds. David Fleay later succeeded in breeding from the young of this first capture and he found that they were summer breeders. Mating, and earliest pouch development, occurs early in November and females lose reproductive condition by mid-January. Teat number seems to be variable and he records that up to twelve newly-born young may be successfully attached.

The Species   Group 29

PIGMY ANTECHINUS, PIGMY MARSUPIAL-MOUSE, *Antechinus maculatus*. All mainland States except Vic.; preferred environment unknown.
Recognition: rather smaller than mouse; head rather cone-shaped in side view; short grey fur; the female, unlike other antechinuses, has a pronounced pouch; tail shorter than head and body.

INGRAM'S PLANIGALE, *Planigale ingrami* (Plate 35). Eastern and northern Qld, N.T., Kimberley and central W.A.; savannah woodland and grassland.
Recognition: much smaller than mouse; head flattened when seen in profile, triangular when seen from above; short grey fur with pepper and salt appearance; tail about equal to head and body.

KIMBERLEY PLANIGALE, *Planigale subtilissima*. Kimberley Division, W.A.; savannah woodland and grassland.
Recognition: much smaller than mouse; externally much like *ingrami*.

NARROW-NOSED PLANIGALE, *Planigale tenuirostris*. North central N.S.W., south central Qld, central W.A.; savannah woodland and grassland.
Recognition: much smaller than mouse; head flattened as in *ingrami* but muzzle rather narrow making head appear less triangular when seen from above; tail longer than head and body in western specimens, about equal in eastern specimens.

Group 30 Dunnarts or narrow-footed
marsupial mice   Plate 36

These soft-furred and delicately built little marsupials are, in general, about the size of a domestic mouse, but a few species are a little bigger. They are small active predators and appear to live principally on insects but, in captivity, small mammals such as mice and baby rats, small birds, lizards, and centipedes are taken avidly, although different species appear to exhibit marked preferences in this respect.

Some species of dunnarts or marsupial mice are fat-tailed in the manner of the Red-eared Antechinus. Like that species, the tail becomes carrot-shaped, or spindle-shaped, in good seasons and (invariably in my experience) when an animal has been kept in captivity for a while. One should be careful when using this character to identify specimens, however, because some individuals of fat-tailed species may have quite thin tails when they are caught in the wild.

Some of the dunnarts have thin tails which never become fat, even under the best of conditions. These thin-tailed species include the largest of the *Sminthopsis* and the rarest of all Marsupials, the Sandhill Dunnart, which is only known from a single specimen. The next rarest is the extraordinary Long-tailed Dunnart of the Marble Bar

35   PLANIGALE, close to *Planigale ingrami*   Group 29

area of Western Australia in which the tail is more than twice the length of the head and body; only four specimens are known of it.

The thin-tailed section of the group includes the Common Dunnart. It is mouse-sized and very soft-furred. It can easily be distinguished from a mouse by its pointed nose and large number of small incisor teeth. When a dunnart is cornered or frightened, unlike a mouse, but like the native-cats, it will adopt a threatening posture with widely open mouth, often accompanied by noisy exhalations.

The fat-tailed members of *Sminthopsis* are, as the need for a fat store would suggest, inhabitants of dry to very dry country. They include the commonest of all dunnarts, the little Fat-tailed Dunnart with its very short, spindle-shaped tail and its alert little foxy face. In northern parts of its range the tail is longer than in the south.

The Fat-tailed Dunnart is the only species of dunnart whose behaviour has been studied in detail. Dr R. F. Ewer has shown that females may start to breed at four months and produce litters continuously at intervals of about eighty-two days for at least six months; it is not yet known whether breeding in the wild is controlled by the seasons or by availability of food. They are cautious and treat new and strange objects appearing in their home ranges with extreme suspicion and, like Sugar Gliders, make extensive use of glandular secretions and saliva as signals (see p. 78). Males fight over females on heat and females, especially older animals when they have pouch young, attack males; sometimes they even kill their mates.

Like bats and some other sorts of small mammal, such as the Pigmy Possum, some species of dunnart can become torpid, or lower their body temperatures well below that needed for normal activity. Dr Gillian Godfrey has shown that the Lara-pinta enters torpor daily in its nest whereas the Fat-tailed Dunnart only does so when food is short. She suggests that torpor may be a state of energy-saving inaction of value when food supplies are unreliable, as they may be in the desert.

The Species   Group 30

SECTION 1: thin-tailed

COMMON DUNNART, COMMON MARSUPIAL-MOUSE, *Sminthopsis murina* (Plate 36). South-west of W.A., south-east of S.A., southern and eastern Vic., eastern N.S.W., eastern Qld; wet and dry sclerophyll, swamps, wet fringes close to rain forest, in hollow logs, black-boys. *Recognition:* as illustrated; foot pads composed of many granules; toe-tips not finely granular.

WHITE-FOOTED DUNNART, WHITE-FOOTED MARSUPIAL-MOUSE, *Sminthopsis leucopus*. Eastern N.S.W., southern and south-eastern Vic., Tas.; sclerophyll forest. *Recognition:* central discs of foot pads finely ridged, not granular.

RED-CHEEKED DUNNART, LUMHOLTZ'S MARSUPIAL-MOUSE, *Sminthopsis rufigenis*. North-eastern Qld, Arnhem Land, north Kimberley; open rocky forest and brushy places in full sunlight. *Recognition:* stripe on top of head; red cheeks.

DALY RIVER DUNNART, *Sminthopsis nitela*. Northern N.T., north Kimberley; tropical woodland, savannah, in burrows or under logs. *Recognition:* rather like *S. murina* but paler; no marked external characters.

LONG-TAILED DUNNART, *Sminthopsis longicaudata*. W.A. (vicinity of Marble Bar), ? Central Australia. *Recognition:* tail more than twice length of head and body, no tuft on end.

SANDHILL DUNNART, *Sminthopsis psammophila*. South-western N.T. (vicinity of Lake Amadeus); Hummock grassland (spinifex i.e. *Triodia*) among sandhills.
Recognition: size much larger than mouse; soles of feet including toes finely granular.

SECTION 2: fat-tailed

FAT-TAILED DUNNART, FAT-TAILED MARSUPIAL-MOUSE, *Sminthopsis crassicaudata* (Plate 37). W.A. (a large area extending inland from the south-west), S.A. (south-eastern extending up into the north-east to the Qld border), south-western Qld, western N.S.W., western Vic.; woodland heaths, grasslands, stony places.
Recognition: tail much shorter than head and body; dark patches on ears and on top of head.

DARLING DOWNS DUNNART, *Sminthopsis macroura*. South central Qld, northern N.S.W.; open plains and woodland.
Recognition: much like *crassicaudata*; but tail much longer, about equal to head and body; colour similar, but dark guard hairs give back more dusky appearance; forehead stripe more distinct extending right forward to nose.

LARAPINTA, STRIPE-FACED DUNNART, *Sminthopsis froggatti*. Northern S.A., western Qld, N.T., central W.A. including southern Kimberley; desert complex, savannah woodland, grasslands.
Recognition: prominent stripe between eyes, extending back between ears; tail as long or longer than head and body.

WHITE-TAILED DUNNART, ASHY-GREY DUNNART, GRANULE-FOOTED SMINTHOPSIS, *Sminthopsis granulipes*. Inland periphery of south-west of W.A.; sclerophyll woodland, to shrub and tree heaths.
Recognition: tail white above and below; fine granules at base of nails on pads of toes instead of smooth toe-tips.

HAIRY-FOOTED DUNNART, *Sminthopsis hirtipes*. The western desert from Charlotte Waters in the east to the Canning Stock Route in the west.
Recognition: pads of feet covered with bristles; very large ears; sharply pointed nose.

Group 31 The Wuhl-wuhl and the Kultarr   Plate 38

The habits of long-legged marsupial mice, close relatives of the *Sminthopsis* group of small carnivorous marsupials, have been misrepresented as long as they have been known. There are two species: the central desert form or Wuhl-wuhl (*Antechinomys spenceri*) and the eastern form or Kultarr (*Antechinomys laniger*).

The first specimen of Kultarr was collected on one of Major Mitchell's expeditions and handed to John Gould who illustrated it. Because the plume-like tail resembles, superficially, that of the Red-tailed Wambenger, Gould drew it in a tree. Subsequently the anatomist Alston disagreed and decided, on anatomical grounds, that the long slender limbs betokened a jumping mode of progression. Unfortunately Alston's statement was misinterpreted to mean that they hopped along like little kangaroos on their hind legs. Thus, until recently, it was believed that these small marsupials were the evolutionary equivalent among marsupials of the Australian hopping-mice (*Notomys*, see p. 144) and they were given the name of 'Jerboa-like Marsupials'.

37　FAT-TAILED DUNNART, *Sminthopsis crassicaudata*　　　　　　　　　　　Group 30

Recently, high-speed photographic studies by myself, and an ingenious method of recording footprints by Basil Marlow, have shown clearly that the normal mode of progression is a graceful gallop in which the animal springs from its hind feet, lands on its fore feet, and then, bringing its hind feet around the outside of its fore feet (rather as though it were leap-frogging), it springs off again with its hind feet. The Plate illustrates some of the postures revealed in the high-speed photography; at the right rear the animal is shown making a springing right-angled turn, while at the left rear, the animal is using its tail as a rudder while it is in mid-air. The standing, two legged posture is that used by the Wuhl-wuhl when it is examining something above the ground.

The body of the Wuhl-wuhl is about the size of that of a large mouse but its legs and tail are much longer.

Very little information is available about the natural history of the Wuhl-wuhl and nothing at all about the Eastern species. The Wuhl-wuhl inhabits burrows, and one dug out by Miss P. Robertson at Warburton Range Mission, in 1963, was in a burrow over five feet in length and ten inches below the surface. Unlike the burrows of most of the native-mice, it appeared only to have one entrance. A specimen captured near Wiluna, Western Australia, was caught down a trap-door spider's burrow. Up to six young have been recorded; there is little information as to breeding season, but the Wuhl-wuhl appears to breed in winter.

The Wuhl-wuhl is a gentle cowering animal, but when it is liberated in strange surroundings it is surprisingly fearless and inquisitive. It is intensely suspicious of holes, cracks, and, when it becomes used to a place, of new objects lying around. The animals which I have kept did not touch meal worms at first, but would eat cockchafer larvae, and large moths and spiders. Lizards placed in a box with them were left untouched.

The Species   Group 31

WUHL-WUHL, PITCHI-PITCHI, *Antechinomys spenceri* (Plate 38). W.A. from the wheat belt to the Pilbara and eastward into north-western S.A., N.T.; woodland, tree steppe, desert associations.
Recognition: brush on end of tail; large ears; delicate slim long legs.

KULTARR, *Antechinomys laniger*. Northern Vic., N.S.W. west of the Divide, southern Qld (and a solitary specimen from a locality on the coast near Cooktown in north-eastern Qld). *Recognition:* slightly smaller and darker than *A. spenceri* but scarcely separable on external features.

Group 32 The Numbat or Banded Anteater   Plate 39

The Numbat is the most beautifully coloured of all the marsupials: its shoulders and general colour are reddish-brown, enlivened by a plentiful sprinkling of white hairs. But across the rump, there are several prominent white bars between which the hair is dark, sometimes almost black. Through the eye there is a prominent dark stripe

38   WUHL-WUHL, *Antechinomys spenceri*

which is framed above and below by long white streaks. The hair of the tail is often carried erect and fluffed out like a bottle brush.

The Numbat has an enormously long tongue with which it collects the termites on which it feeds. John Calaby has shown that in south-western Australia it is commonly found in Wandoo forest (*Eucalyptus wandoo* or *Eucalyptus redunca* var. *elata*) in which a high proportion of mature trees are even attacked by termites while living; in addition, the forest floor is littered with fallen branches. These are hollow and provide both shelter for the Numbat and food for termites on which the Numbat feeds.

Unlike most marsupials, the Numbat is around during daylight hours and a good deal of its time is spent in searching for food. It walks slowly along, smelling the ground here and there. When termites are located in sub-surface soil, it digs rapidly, licking up the insects as they are exposed. Numbats are remarkably unafraid and they can be watched from a car in State forests within an easy day's run from Perth.

There are usually four young born between January and April, or May, which are carried or nursed by the mother through winter. There is good evidence that the female Numbat digs a burrow and on several occasions young have been found in nests at the end of these short breeding-burrows. At other times, the Numbat seems to live in hollow logs. If cornered in a hollow log, the Numbat makes a low throaty growl or a noise like escaping steam.

Today the main population of Numbats seems to be confined to the Wandoo-bearing country of the south-west of Western Australia; in the past, however, it extended farther east as far as Kalgoorlie. East of this the Numbat appears to have occurred in a thin but probably continuous population through Laverton, the Warburton Range, and through to the Everard Range, in northern South Australia. There it is a redder animal than it is in the south-west. In the early days of zoological exploration it even occurred as far east as south-western New South Wales.

The Species   Group 32

NUMBAT, BANDED ANTEATER, *Myrmecobius fasciatus* (Plate 39). Southern W.A., to north-western S.A., formerly south-western N.S.W.; sclerophyll forest and woodland (its habitat in the desert areas has not been recorded).
Recognition: transverse white and nearly black bars on reddish-brown background; long-haired tail; total length about eighteen inches.

Group 33 The Thylacine   Plate 40

The Thylacine is the largest of all the marsupial predators of Australia and because of its sheep-killing habits was persecuted to the brink of extinction by 1914. Dr Eric Guiler, who has done the most recent work on the status of the Thylacine in Tasmania believes that it is possible that it is slowly recovering its numbers today.

At first sight the Thylacine looks to be rather dog-like but, as is shown in the Plate on p. 131, the hind quarters and base of the tail are curiously tapered and there are prominent transverse stripes across the lower back and hind quarters. The two postures

in the Plate are based upon photographs and film of the last animal to be held captive in the Hobart Zoo; it died in 1933 (see p. 201). In the foreground the animal is rearing up as the keeper rattles the bars of its pen. There is no evidence for the old story that the Thylacine was able to jump along on its hind legs like a kangaroo.

Dr Guiler says that mating of Thylacines in Tasmania mostly took place in the summer, and three or four young were born. He points out that there was some breeding throughout the year as well because young have been found in the pouch at all times.

Bounties were paid by the Van Diemen's Land Company as early as 1840, at what was then a high figure of six shillings per scalp for less than ten scalps, and eight shillings each for ten to twenty, and ten shillings each for more than twenty scalps. This suggests that the destruction of Thylacines was of considerable economic importance to the farming community. A total of 2,184 Thylacines were killed and presented for government bounties between the years 1888 and 1909. Dr Guiler emphasizes that this total is substantially below the number of Thylacines actually killed because he has been assured by old trappers that about half of those killed were taken round properties and larger bounties collected from private owners. The numbers taken, as shown in the official figures, suddenly declined in 1905.

The distribution of bounty payments within Tasmania also indicates that the Thylacine was widely distributed throughout the State except in the south-west where no bounties were claimed. This is despite the fact that during the bounty period there were more men working in the south-west of Tasmania than there are today. As a result, Dr Guiler has concluded that Thylacines were never plentiful in the area of high rainfall, rain forest and buttongrass plains. Recent indications, however, point to the fact that small numbers of Thylacines may have persisted in the south-west. A case of sheep-killing in the Derwent Valley near Hobart in 1957 is believed by Dr Guiler to be due to a Thylacine.

It is generally accepted by zoologists that the Thylacine, like the Tasmanian Devil, had been absent from the Australian mainland for many thousands of years. There are numbers of reports, however, spread over many years, of sightings of Thylacine-like animals on the mainland, from places as widely separated as north-eastern Queensland, western New South Wales and the southern Nullarbor in Western Australia, but none have been confirmed. There is no doubt that the Thylacine was widely distributed throughout Australia and even New Guinea during the last ten thousand years but the most recent accurately dated mainland remains are of a carcass found by Mr and Mrs David Lowry in a cave near Eucla on the edge of the Nullarbor. The carcass which has recently been dated by radio-carbon is of an animal which died between 2940 and 2240 B.C.; it was lying on the surface. The specimen is completely mummified and possesses a full coat of hair which displays the characteristic stripes. The soft tissues are preserved and dried.

The Species   Group 33

THYLACINE, TASMANIAN WOLF, TASMANIAN TIGER, *Thylacinus cynocephalus* (Plate 40). Tas.; savannah woodland, or open sclerophyll forest, with rocky outcrops near by.
Recognition: dog-sized; olive-brown with darker transverse stripes across hind quarters; tail merging widely into hind quarters.

40 THYLACINE, *Thylacinus cynocephalus*   Group 33

Group 34 The Marsupial-mole   Plate 41

From the evolutionary point of view, the Marsupial-mole is one of the most spectacular of all marsupials. There is no doubt that it is a marsupial—its pouch reveals
this—otherwise it resembles the true moles of other lands in the form of its body
and its scoop-like feet.

The Marsupial-mole is a small animal, only about six inches in length; it has
beautiful fine silky fur which varies in colour from almost white to a rich golden
orange. When the animal is alive, it has a remarkably velvety, almost iridescent,
sheen. There is a horny shield covering the nose, no eyes are visible, and there is no
external ear beyond a small hole. The tail is a strange little bare, ringed, cone.

Little is known about the Marsupial-mole in the wild and what accounts there are
are merely observations of what individuals were doing at the time of their capture.
These all seem to have been found on the surface of sandhills in sandridge desert
country. The Mole is not known to construct permanent tunnels. Sir Edward Stirling,
quoting informants who had seen them, said that the sand falls in behind the burrowing
Mole, as it goes. While it is underground, its progress can be detected by a slight
cracking, or moving, of the surface over its position. When it emerges at intervals to
travel on the surface it moves at a slowish pace with a peculiar sinuous motion. The
belly is flattened against the soil and the fore limbs are doubled beneath it so that the
animal rests on their outsides. Propulsion on the surface seems to be mainly by the
hind feet and its track is a peculiar wavy triple line; the central furrow of the track is
continuous, being made by the tail. The others are broken and are made by the
hind feet.

Professor Wood Jones has provided us with useful notes of the living Mole in
captivity; these were made in 1923. He wrote that the captive displayed the most
remarkable nervous activity, making endless tours around its box with characteristic
energy and haste. It feverishly searched the corners of its cage and on discovering
something edible, such as a handful of earthworms, it would noisily account for it
with a maximum of speed. Having finished, it would start its hasty touring again
until, suddenly, without warning, it would fall asleep. Wood Jones said that even
its sleep seemed hurried with rapid breathing and sudden awakening, immediately
followed by rapid movement. He described its gait as a rapid shuffle and the movement of its body as fluid and sinuous so that, when held in the hand, it seemed to
flow through the fingers.

Nothing is known of breeding beyond that Wood Jones found one young in a pouch.

The Species   Group 34

MARSUPIAL-MOLE, *Notoryctes typhlops* (Plate 41). The Western Desert from Ooldea on the
Transcontinental Line in the south, Charlotte Waters in the east, and Wollal north of
Port Hedland in the west; sandridge desert with acacias and shrubs.
Recognition: larger than mouse; horny shield on nose; no eyes; bare, stubby, ringed tail.

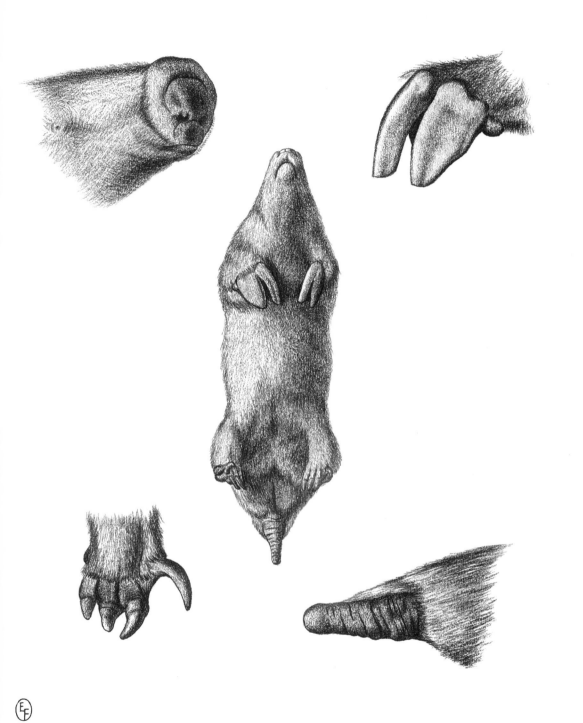

41   MARSUPIAL-MOLE, *Notoryctes typhlops*   Group 34

9

Australian native rats and mice

LACKING the glamour of the pouched mammals, and, moreover, being naturally inconspicuous, the native rats and mice are the most poorly known and least understood of all our mammals. Many species are only recorded by a mere handful of skins in museum collections, and nothing is known of the habits of many of them.

The native rats and mice fall into four very distinct groups. The first group comprises the species of cosmopolitan genus *Rattus* which occurs through most of the world; most of the species which occur in Australia are confined to this continent, although some of them also occur in New Guinea and the other islands to the north. This group can be called our Native Bush-rats; all its members are true bush-dwellers, secretive, and seldom seen by people despite the fact that they are the close relatives of the introduced Rat, which is so common around our houses and gardens, and is responsible for the little caches of emptied-out snail shells found in garages and garden sheds. This rat (*Rattus rattus*) is often called the Ship Rat or the Black Rat—names which convey neither the habits nor the colour of the rat. The species is found far from ships, while the colour varieties found wild in Australia include black, blue-grey, brown and white-bellied brown animals.

The second group of Australian native rodents are the water rats, of which there are two main Australian kinds—the large Water Rat or Beaver Rat, and the much smaller and very rare False Swamp-rat—only known from seven specimens. These are Australian representatives of a group which is centred in New Guinea.

Next there are those very characteristically Australian members of the Rat and Mouse family which are called by zoologists the 'Old Endemics'. These include the most spectacular of all the Australian native-mice and have no really close relatives anywhere else. Among them are included such forms as the tree-rats, the hopping-mice, stick-nest rats, the rock-rats, and a group of more conventionally-appearing native mice which belong to the genus *Pseudomys*.

Finally, the fourth group comprises a number of species of rats—largely arboreal—which, like the water rats, have a centre of distribution in New Guinea. These are the mosaic-tailed rats which are characterized by the mosaic-like arrangement of the scales of their tails.

The classification of the native rats and mice into species is still a matter of some controversy—largely because adequate biological information is not yet available for most of the groups which would enable species to be separated on grounds other than physical characteristics. For many years now Mr J. A. Mahoney of the University of Sydney has been working on this problem; I have discussed many of the known specimens with him and agree with his interpretation of them. The arrangement of the native rats and mice presented here is the result of the conclusions reached by him after morphological studies, supplemented by biological data when available, on the major collections of Australian rodents in the museums of Australia and Europe.

134

The identification of rats and mice is much more difficult than that of other main groups of Australian mammals. This is because most of the species are structurally close variations on a single theme; their closeness is probably an expression of the fact that they have only recently radiated within Australia—that is, recently by comparison with the times of radiation of the other Australian principal groups (see p. 34). The external characters used for identification include comparison between lengths of head and body and the tail, the length of the ear, the quality and to a lesser extent the colour of the body fur, the extent of hairiness or tuftedness of the tail and its colour, and the distribution of the nipples. The nipples or mammae of rats and mice are arranged in two major areas; one set, the pectorals, are arranged on the chest (sometimes extending outwards in a crescent behind the armpits); the other set, the inguinals, are arranged in the groin. The maximum number is three pectoral and three inguinal pairs (i.e. twelve teats altogether); the least number is no pectoral and two inguinal pairs. Unfortunately, teats can only be seen well in adult females.

For definite identification it is usually necessary to examine the characters of skull and teeth, but this is a difficult procedure and not one to describe in this book. External characters combined with knowledge of locality and habitat are usually sufficient to obtain identification to within a few alternatives.

Group 35 Bush-rats   Plate 42

Much biological information must yet be gathered before we can be fully certain of the species of Bush-rats, and it can only be hoped that those forms among them which are little known can still be relocated to enable such knowledge to be obtained. An excellent example of what is needed is given by studies by Dr Mary Taylor and Dr Elizabeth Horner on the populations of the Southern Bush-rat, *Rattus fuscipes*. They have collected large numbers from various places and bred them, in captivity, in order to detect the different species by their inability (or reduced ability) to inter-breed (see p. 28). As a result, we now know that three supposed species (*Rattus fuscipes*, *Rattus assimilis* and *Rattus greyi*) are no more than geographically separate populations of a single southern species which we can call the Southern Bush-rat (*Rattus fuscipes*). In addition to clarifying the relationships of species, Dr Horner and Dr Taylor have also revealed much information of breeding biology in native rats.

As well as being difficult to identify, the species of *Rattus* are poorly known. They occupy a number of different habitats, but it is notable that only two of them occur in the Centre. One is the Long-haired Rat whose populations appear to undergo cyclic increases at from five to seven-year intervals; at these times it swarms from western Queensland into adjoining parts of eastern Central Australia. These migratory populations may persist for as long as eighteen months, but may vanish quite quickly at other times. Finlayson has recorded that in non-swarming years there are small populations at low density in parts of the Centre. The other species which occurs in the Centre is a form of Tunney's Rat which, except for its occurrence there and in western New South Wales, is otherwise a coastal species. It was collected in the vicinity of Alice Springs and at Tennant Creek in 1894; twelve specimens of it are known from the Centre.

The two species of native *Rattus* most likely to be known to Australians are the Southern Bush-rat and the Eastern Swamp-rat. The Southern Bush-rat can be found in many habitats ranging from rain forest to coastal heaths, but it seems to prefer the dense vegetation of moist places. It breeds more or less all the year round, although breeding activity reaches a peak in autumn and decreases after that for three months. The gestation period is between twenty-two and twenty-four days and this may be prolonged in lactating females; it is suspected by Dr Taylor that this Bush-rat may, in a manner reminiscent of some kangaroos and wallabies (see pp. 42), be able to hold the fertilized egg in its uterus for a while in arrested development. The young, of which there are normally three or four, are weaned by about twenty-six days and it seems likely that the life-span does not exceed more than one year in the natural habitat. During that time the average female produces three or four litters.

The Eastern Swamp-rat has been most recently studied by Mr R. H. Green of the Queen Victoria Museum, Launceston. He found that, in Tasmania, the species occurs in a wide range of habitats from coastal swamp land to rain forest. It makes tunnels in dense undergrowth, around swamps, for example, and has no objection to wet and muddy conditions underfoot; but it seems to prefer to keep its body fur dry. He found no evidence, in stomach contents, that the Swamp-rat fed in water, and showed, moreover, that, provided green feed is available, it does not need to drink in order to maintain good condition.

Although the swamp-rat does not burrow in the water-logged soil around swamps, when light sandy soils are available under sufficiently dry conditions it is a burrower and constructs shallow warrens occupying as much as a hundred square yards. In the wild in Tasmania the species has a restricted summer breeding season; litters range in size from three to six, and two litters may be produced in a season. The young are born naked and are weaned at about twenty-five days. Most individuals live a little more than a year, but very few complete two full years of life. In captivity, breeding is not restricted to a narrow season and twenty-seven young have been born to one captive pair in eleven months.

The Species   Group 35

For the recognition of introduced rats see the end of the list of species.

SOUTHERN BUSH-RAT, *Rattus fuscipes* (Plate 42). Eastern Qld (south of Cairns), eastern N.S.W., Vic, coastal S.A. as far west as the Eyre peninsula, Kangaroo Island and other islands off S.A., south-western W.A., Recherche Archipelago and Houtman Abrolhos, W.A.; rain forest floor, sclerophyll (especially moist gullies), coastal complex especially swamps; prefers moist situations but can even be found on sandy heaths.
Recognition: brown, long guard hairs giving it a fluffy look; tail about the same or shorter than head and body. Females with 2 pectoral and 3 inguinal pairs of mammae except in northern Queensland where they have 1 pectoral and 3, or sometimes 2, inguinal pairs.

EASTERN SWAMP-RAT, *Rattus lutreolus*. Coastal S.A. south-east of Adelaide, coastal Vic., Tas., Flinders Island, eastern N.S.W., eastern Qld; rain forest, wet sclerophyll, swamps, other wet grasslands and heaths.
Recognition: dark brown above, belly dark grey; tail very much shorter than head and body. Females with 2 pairs of pectoral and 3 inguinal pairs of mammae except in Tasmania where they have 2 pectoral and 2 inguinal pairs.

42 SOUTHERN BUSH-RAT, *Rattus fuscipes*

DUSKY FIELD-RAT, NORTHERN BUSH-RAT, CANEFIELD-RAT, *Rattus sordidus.* Eastern Qld from Darling Downs northwards, northern N.T.; woodland, tropical savannah woodland, canefields, river flats, ricefields.
Recognition: fur rather harsh, some populations with long guard hairs; dark brown with light grey-brown bellies; tail a little shorter than head and body. Females with 3 pectoral and 3 inguinal pairs of mammae.

LONG-HAIRED RAT, *Rattus villosissimus.* Western Qld, N.S.W. west of Great Dividing Range, north-eastern S.A., eastern N.T.; grasslands.
Recognition: very long guard hairs, especially on rump. Pepper and salt colour (greyish and sandy), whitish belly; tail about same length as head and body. Females with 3 pectoral and 3 inguinal pairs of mammae.

MOTTLE-TAILED CAPE YORK RAT, *Rattus leucopus.* Eastern Cape York Peninsula as far south as Cairns; rain forest.
Recognition: feet very white, belly white or yellow; tail about same length as head and body, often with whitish blotches. Females with 1 pectoral and 2 inguinal pairs of mammae.

TUNNEY'S RAT, PALER FIELD-RAT, *Rattus tunneyi.* Coastal W.A. northwards from Moore River, Kimberley, northern N.T., Melville Island, eastern Qld south to Rockhampton, western N.S.W., Central Australia (Alice Springs, Tennant Creek); sandy flats behind beach lines, river flats with friable soil, grassland, heaths, savannah woodland.
Recognition: sleek spiny yellowish-brown or greyish hair, white or yellow-bellied; tail shorter than head and body. Females with 2 pectoral and 3 inguinal pairs of mammae.

Rattus manicatus and *R. exulans* are omitted from the list of native species. The former is possibly based upon a single mislocalized specimen of the Southern Bush-rat, *R. fuscipes*; and, although *R. exulans* has been recorded from Adele Island, Western Australia, it is likely to be a human introduction. It is a widespread Pacific species, often commensal with man.

RECOGNITION OF INTRODUCED RATS

RAT, BLACK RAT, SHIP RAT, *Rattus rattus*
Recognition: colour variable from grey-bellied black, through brown with greyish-brown belly, to white-bellied brown; tail longer than head and body. Females with 2 pectoral and 3 inguinal pairs of mammae.

NORWAY RAT, BROWN RAT, SEWER RAT, *Rattus norvegicus*
Recognition: greyish-brown or yellowish-brown above, pale greyish below; tail obviously shorter than head and body. Females with 3 pectoral and 3 inguinal pairs of mammae.

PACIFIC ISLANDS RAT, PACIFIC RAT, *Rattus exulans*
Recognition: small, brownish, rather spiny guard-hairs; tail about same length as head and body. Females with 2 pectoral and 2 inguinal pairs of mammae.

Group 36 The Water Rat and the False Swamp-rat   Plate 43

The Water Rat is, apart from the Platypus, the only highly specialized fresh-water mammal in Australia. It is entirely carnivorous, living on fish, fresh-water crayfish, mussels, and even birds, rats and lizards. Fully grown adults may be two feet in length and weigh in the vicinity of two pounds. They appear to have poor vision.

43   WATER RAT, *Hydromys chrysogaster*

The way in which the Water Rat opens mussels and extracts their contents without breaking the shells has often been argued about. Recent observations by Mr G. J. Barrow of the Innisfail Field Station of the Queensland Institute of Medical Research have now shown that Water Rats collect the mussels and put them in the sun until the heat of the day causes the shells to open.

Water Rats have up to four young which, by the time they are thirty-four days old, are capable of independent existence. David Fleay, who has bred them in captivity, has found that a female will bear successive litters at two-monthly intervals. He is of the opinion that some three litters (varying from two to four in number) are produced during spring and summer. The Water Rat seems to increase slowly, and, if its populations are to be cropped commercially for the fur trade on a continuing basis, considerable control will have to be maintained. The fur is extremely beautiful and much in demand in those States in which they are nowadays permitted to be taken (New South Wales, South Australia, Tasmania and Queensland). They are rather variable in colour; the south-western population is a rich glossy black, that of the south-east varies from individuals with golden-brown backs and orange bellies, to grey with white under-parts. Individuals from northern Australia are usually a greyish-brown.

The much smaller False Swamp-rat of northern Queensland and the Northern Territory is one of the rarest of our native mammals—nothing has been published about its biology. Unlike the Water Rat it does not have webbed hind feet. The Water Rat and the False Swamp-rat belong to a sub-family of rats distinct from all other Australian rats. A number of species occurs in New Guinea and the islands to the north of Australia, and they are distinguishable from the other rats and mice by possessing fewer molar teeth. Both Australian species possess two pairs of molars on each side of the jaw as compared with the three usual in rats and mice.

The Species   Group 36

WATER RAT, BEAVER RAT, *Hydromys chrysogaster* (Plate 43). Fresh-water rivers and creeks of Australia. Sea beaches on some islands (e.g. Barrow I., W.A., Magnetic I., Qld, islands in Bass Strait).
Recognition: much larger than ordinary rat; characteristic shape (see illustration); fur very dense and soft; hind feet webbed; tail-tip white. Females with 2 pairs of mammae only, both inguinal.

FALSE SWAMP-RAT, *Xeromys myoides*. Qld (Mackay), N.T. (Alligator River); habitat at Mackay was a permanently-filled swamp covered with tall grass, shrubs and pandanus.
Recognition: size of small rat; head long and *Hydromys*-like, ears short and round, feet not webbed; colour dark slaty-grey, under-surface white; tail much shorter than head and body. Females with 2 pairs of mammae only, both inguinal.

Group 37 Tree-rats   Plate 44

The Golden-backed Tree-rat is one of the most beautiful and agile of our native rodents. So little was known about the habits of this and the closely related Black-footed Tree-rat that they were thought to be some kind of giant hopping mouse. It is

44   GOLDEN-BACKED TREE-RAT, *Mesembriomys macrurus*   Group 37

now known that these two species of *Mesembriomys* and the other tree-rats, *Conilurus*, are climbers, descending to the ground to feed.

The Golden-backed Tree-rat is very rare in collections but it is not uncommon today in north-western and northern Australia from West Kimberley through into the Northern Territory. It was originally collected in 1875 from near Roebourne in Western Australia, but no specimen has since been taken from south of Broome in the Kimberley.

The Black-footed Tree-rat is to be found in tree-covered country in northern Australia. It also occurs on Melville and Bathurst Islands. The American/Australian Scientific Expedition to Arnhem Land showed that it still occurred there; they found that it made nests of leaves and bark within hollow trees. They also found that Black-footed Tree-rats had established themselves in the temporary buildings of a wartime airstrip at Gove Airfield in Cape Arnhem Peninsula. Mr K. Keith of the Division of Wildlife Research of C.S.I.R.O. has also established that it is common around Port Essington on the Cobourg Peninsula. All naturalists who have handled the Black-footed Tree-rat have noted its savage temper and the severity of its bite, which is not surprising since it is powerful and agile and has sharp chisel-like incisors. When enraged it is loud-voiced; Finlayson said of a captive that it would raise its voice progressively into a sort of whirring machine-like crescendo, not unlike the noise made by a Sugar Glider (see p. 79).

The Brush-tailed Tree-rat is still common in the Northern Territory and, when surprised on the ground, flees to the nearest tree, which it climbs.

Little is known of the habits of the White-footed Tree-rat of eastern Australia; early settlers called this the Rabbit Rat because of its rounded form and long ears. It has not been seen alive in this century. It nested in hollow limbs.

The Species Group 37

Females have 0 pectoral and 2 inguinal pairs of mammae.

GOLDEN-BACKED TREE-RAT, *Mesembriomys macrurus* (Plate 44). W.A. north of Fortescue, northern N.T.; woodland savannah.
Recognition: form as illustrated, tail-tip white, feet white, orange-brown patch on back, fur harsh.

BLACK-FOOTED TREE-RAT, *Mesembriomys gouldii.* North Kimberley, coastal N.T., Melville Island, Bathurst Island, Cape York Peninsula (north of the Atherton Tableland); woodland savannah.
Recognition: form as above; black ears and feet, but digits may be white; long black guard-hairs on rump; fur harsh.

BRUSH-TAILED TREE-RAT, *Conilurus penicillatus.* Northern Kimberley, northern N.T., Melville Island; savannah woodland.
Recognition: form similar; tail-tip usually black or occasionally white; when white, portion between it and the body is black; feet white; fur harsh.

WHITE-FOOTED TREE-RAT, RABBIT RAT, *Conilurus albipes.* South Qld., N.S.W., S.A.
Recognition: form similar; tail-tip white with white extending along under-side of tail to body; feet white; fur long, close and soft.

Group 38 Stick-nest rats

Plate 45

The stick-nest rats or house-building rats are gregarious, fluffy-furred, short-faced, little animals which, like some of the American desert rodents, build houses of sticks and, presumably, like those nest-builders, use their houses as a device to enable them to withstand the desiccating heat of the desert days.

There are two mainland species, the Stick-nest Rat and the White-tipped Stick-nest Rat. The stick-nest rats which occur abundantly on Franklin Island, in the Nuyts Archipelago off South Australia, probably belong to the same species as the Stick-nest Rat of the mainland; they are not listed separately from it here.

From descriptions of houses of stick-nest rats in different situations it seems likely that terrain and availability of building materials may play a large part in determining the form of the houses. When they build houses on the open plain, the Stick-nest Rats interweave sticks from six to twelve inches in length, around, and among, the lower branches of a stunted bush, building compact structures of three feet in height. There are numbers of entrance holes around the periphery, and it seems that there is usually a shallow burrow beneath the mass. In the absence of bushes to serve as a foundation, nests become flattened, loose, heaps of sticks. Mr Ellis Troughton, at Fisher on the Transcontinental Line in 1922, found that these are often associated with rabbit warrens which the rats use. They also have stones incorporated into them. Heaps of stones and sticks covering deep crevices, and showing signs of occupation by a large rodent, occur under the overhangs of rocky breakaways in the Western Desert; it is likely that these are houses of the Stick-nest Rat made in a more sheltered position than on the plain.

Professor Wood Jones, the discoverer of the Stick-nest Rat of Franklin Island, says that these gentle rats do not burrow themselves. On the island they often place their nests, which may consist of no more than a few sticks, or seadrift, in relation to a hole such as a cleft in rocks. When large nests are constructed out on the flats, they may be placed over the burrows of penguins. He noted, too, that the rats were independent of fresh water and that their staple article of diet was the succulent *Tetragona implexicana* of which huge quantities were consumed. He found that he was unable to attract them into traps with any bait. They are gentle and tame, even when freshly caught, and are reminiscent of little rabbits with long tails.

There is no good modern account of the White-tipped Stick-nest Rat, but Gerard Krefft, who observed it on the Murray and Darling Rivers in 1857, said that it occupied the homes of the Stick-nest Rat after they had been abandoned. However, there seems little doubt that the White-tipped species does build its own nest because Mr H. H. Finlayson has pointed out that it has been collected from nests some hundreds of miles away from the nearest records of the other species.

Like other members of the 'Old Endemic' rodents, these rats drag their young around attached to their teats. Adults are about a foot in length from nose-tip to tail-tip.

The Stick-nest Rat will also build nests in captivity if supplied with suitable material. Captives kept at the South Australian Museum appeared to arrange sticks with great care; the rats pulled them into final position from within the nest.

The Species   Group 38

Females have 0 pectoral and 2 inguinal pairs of mammae.

STICK-NEST RAT, HOUSE-BUILDING RAT, *Leporillus conditor* (Plate 45). Northern and central S.A. and western N.S.W. (from the eastern part of the Nullarbor to the Darling R.), Franklin I., Nuyts Archipelago, S.A. (possibly occurred in the Victorian Mallee along the south bank of the Murray R.); sclerophyll woodland, shrub and tree heaths, salt-bush plain.
Recognition: size larger than rat; tail not white-tipped, untufted; tail shorter than head and body.

WHITE-TIPPED STICK-NEST RAT, TILLIKIN, *Leporillus apicalis.* Central Australia (from as far west as the Mann and Musgrave Ranges), S.A., to Western N.S.W., Vic. (along the Murray from about Euston westward); sclerophyll woodland, shrub and tree heaths.
Recognition: size of rat; tail tipped with white, untufted; tail longer than head and body.

Group 39 Hopping-mice   Plate 46

The Australian hopping-mice have become adapted for rapid bi-pedal, or ricochetal, locomotion in the same way as kangaroos. Outside Australia, several unrelated groups of rodents have achieved similar habits under desert conditions. They are widespread in arid and northern parts of Australia but a few, such as Mitchell's Hopping-mouse, occur in more densely inhabited country in the south.

A glandular specialization of unknown function occurs in all* species of the hopping-mice. It is a patch of specialized tissue upon the throat (the gular gland), or on the chest (the sternal gland), which may even take the form of a recessed pocket or pouch. This organ contains numerous glandular openings and specialized hairs. The White-striped Mastiff Bat (see page 168) and some other bats have a similar structure, and, in that case, too, its function is quite unknown.

As in the case of many desert rodents the hopping-mice are subject to great fluctuations in numbers; when they are common, they are very common indeed. Following good seasons, localities may be swarming with them and great numbers may be seen from a car, crossing the road in the headlight beam; at other times, the same localities may appear to be completely deserted, and mice absent.

They are nocturnal and, although they will drink water readily when it is available, experimental work by Dr Richard E. MacMillen and Dr A. K. Lee on two of the desert species, *Notomys alexis* and *N. cervinus*, has shown that they are capable of maintaining themselves on a diet of mixed birdseed without any water at all. In fact, they even gain weight. They survive through their remarkable ability to excrete their waste nitrogen in the form of the most highly concentrated urine yet known in any rodents, and through their behaviour as well. They avoid extreme heat and desiccation by remaining underground during the day.

Hopping-mice are efficient burrowers; they build complex burrow systems with numbers of entrances, many of them vertical pop-holes. They block up the holes from inside when they are within.

*External characters are not yet known of one species, i.e. the Darling Downs Hopping-mouse, *Notomys mordax.*

45   STICK-NEST RAT, *Leporillus conditor*   Group 38

The arrangement into species is still much in need of confirmation with biological data. Nineteen different names have been proposed at various times, but it seems, from careful examination of specimens by Mr J. A. Mahoney, that there are not more than nine species—and there may even be less when we come to know more about the biology of the forms which are only known today from a few specimens.

The Species   Group 39

Females have 0 pectoral and 2 inguinal pairs of mammae.

MITCHELL'S HOPPING-MOUSE, *Notomys mitchellii* (Plate 46). Western N.S.W., north-western Vic., S.A., W.A. (as far north as Shark Bay, and excluding high rainfall areas of the south-west); sclerophyll woodland, tree and shrub heaths, grasslands.
Recognition: body-size that of large mouse; no separate gular and sternal glandular areas or gular pouch, but area marked by shining white or creamy hairs extending from throat to sternum; incisors not grooved.

SPINIFEX HOPPING-MOUSE, DARGAWARRA, BROWN HOPPING-MOUSE, *Notomys alexis*. Western Qld, N.T., northern S.A., W.A. (in about latitudes 20°–26°S.); desert complex, dunes, grasslands (including spinifex—*Triodia*).
Recognition: body-size as above; separate gular pouch and sternal glandular areas (marked by shining white hair) in males, gular pouch in females; incisors not grooved.

DUSKY HOPPING-MOUSE, WILKINTIE, WOOD JONES' HOPPING-MOUSE, *Notomys fuscus*. Between Rawlinna, W.A., Ooldea, south-western S.A., and Birdsville in south-western Qld.
Recognition: body-size as above; gular pouch very large and deep (an obvious pocket) with mass of shining white hairs within it; incisors not grooved.

FAWN-COLOURED HOPPING-MOUSE, OORARRIE, *Notomys cervinus*. Between Ooldea, south western S.A., Charlotte Waters, southern N.T., and Birdsville in south-western Qld.
Recognition: body-size as above; no gular gland or pouch, sternal glandular area between fore legs marked by white hair; upper incisors grooved along their outer surfaces.

BIG-EARED HOPPING-MOUSE, *Notomys megalotis*. W.A. (vicinity of New Norcia).
Recognition: known only from material in poor condition, external features not discernible except a little larger than *cervinus*; upper incisors grooved along their outer surface.

LONG-TAILED HOPPING-MOUSE, *Notomys longicaudatus*. W.A. (vicinity of New Norcia), southern N.T., north-western N.S.W.
Recognition: body-size of domestic rat; quite small ears, much shorter than head length; no gular gland or pouch, males with a sternal glandular area between fore legs; incisors not grooved.

SHORT-TAILED HOPPING-MOUSE, BRAZENOR'S HOPPING-MOUSE, *Notomys amplus*. N.T. (vicinity of Charlotte Waters).
Recognition: body-size of domestic rat; very large ears nearly as long as head; gular glandular area well marked; incisors not grooved.

NORTHERN HOPPING-MOUSE, *Notomys aquilo*. N.T. (Groote Eylandt), western Cape York Peninsula; coastal dunes.
Recognition: body-size of large mouse; not distinguishable externally from *alexis*.

DARLING DOWNS HOPPING-MOUSE, *Notomys mordax*. Qld (Darling Downs).
Recognition: External characters unknown.

46 MITCHELL'S HOPPING-MOUSE, *Notomys mitchellii*

Group 39

Group 40 Rock-rats

Plate 47

Three closely related species of native rodent are always found among central and northern rocky outcrops. Adults of these rock-rats are easily recognized because they have thick tails rather like some species of marsupials. In juveniles the tail may be scarcely enlarged, but in adults the thickened, strongly ringed, and rather heavy tails are usually very obvious. They are also very fragile. Although other rats and mice lose their tails with ease if they are maltreated, the hair and flesh strip from the vertebrae of rock-rats particularly easily; the animal soon amputates the remaining naked skeleton to leave a shortened stump.

Rock-rats, like most 'Old Endemics' (see p. 134), drag their young around attached to their teats.

The Species

Females have 0 pectoral and 2 inguinal pairs of mammae.

COMMON ROCK-RAT, *Zyzomys argurus* (Plate 47). W.A. (Pilbara and Kimberley), N.T.*, north Qld; rocky outcrops.
Recognition: about size of small rat; fine, slightly stiff fur, yellowish-brown above, pale below; tail sparsely haired except towards end where long hairs extend beyond tip; tail usually thickened in adults, sometimes broken off short.

WOODWARD'S ROCK-RAT, LARGE ROCK-RAT, *Zyzomys woodwardi*. W.A. (Kimberley), N.T. (Arnhem Land); rocky outcrops.
Recognition: body larger than that of domestic rat but chunky (vole-like or guinea-pig-like); fur crisp and harsh, grizzled, paler below; tail thickened, well-haired, often broken off short.

MACDONNELL RANGE ROCK-RAT, *Zyzomys pedunculatus*. N.T. (MacDonnell and James Ranges); presumably rocky outcrops.
Recognition: size of domestic rat; fur soft, brown; tail thickened and well furred (as hairy as back); tail pale below but same colour as back above.

Group 41 The Broad-toothed Rat

Plate 48

From fossil evidence we know that the Broad-toothed Rat was once much more widely spread through southern and south-eastern Australia than it is today. Nowadays, it appears to be a relict species, which only survives as isolated colonies in places in New South Wales, Victoria, and Tasmania which provide it with the cold, humid, or alpine, conditions that it requires.

*The first specimen of the Common Rock-rat collected was said to be from South Australia, but at a time when the Northern Territory was part of South Australia. It has not been collected since in South Australia.

47   COMMON ROCK-RAT, *Zyzomys argurus*

Group 40

The most comprehensive study of the Broad-toothed Rat in mainland Australia was made by Mr John Calaby, of the Division of Wildlife Research, and Mr D. J. Wimbush of the Division of Plant Industry, C.S.I.R.O. The population studied by them was in Kosciusko State Park where the rat lives along small creeks among shrubs and long grass.

There, the climate is rigorous; annual precipitation is of the order of seventy-five to ninety inches, much of it falling as winter snow. The mean temperature for the hottest month is between 50° and 55° Fahrenheit, and the mean air temperature for the three coldest months seldom goes above freezing. The vegetation among which the rodent lives is a heath community of shrubs and long grass which is covered by snow during the winter. Climatic conditions beneath the snow are relatively mild and the rat lives in tunnels among the shrubs. By contrast, Victorian colonies live at low altitudes; although wet they are less cold.

In Tasmania, the best known population of the Broad-toothed Rat is in Cradle Valley, where it was discovered by Mr H. H. Finlayson sixty-one years after the previously last known museum specimen had been collected. This colony is at about 3,000 feet and also has a severe climate although the snow cover does not appear to persist through the winter. Subsequent work by Mr R. H. Green on the Broad-toothed Rat in Tasmania has shown that it is apparently limited in its distribution to the central-western, western, and south-western parts of the island; from altitudes near sea-level at Port Davey to about 3,000 feet near Cradle Mountain, as already mentioned. It is very localized and is limited to a single type of habitat; this consists of vegetation on very boggy ground such as that of the drainage systems of wet sedge-lands in openings in rain forest, or button grass or tussock grass areas. It is invariably found in association with the Eastern Swamp-rat (*Rattus lutreolus*) and the Swamp Antechinus (*Antechinus minimus*). While snow lies in their alpine habitat, the rats do not leave the shelter of their tunnels through the vegetation. The mean annual rainfall of the Tasmanian habitat of the Broad-Toothed Rat is about 100 inches.

It is probable that breeding in the wild Kosciusko colony begins in the spring and litters are produced during the summer. Litters of up to three have been recorded; the young are precocious, and are large and well furred at birth, although their eyes are closed. Like most of the 'Old Endemics', the young attach themselves tenaciously to the teats and are dragged around on their backs by their mother. At three weeks of age the young are large and active but still hang on to the teats and run behind the mother. They probably reach adult weight at about sixteen weeks.

In captivity, within a day or so of the birth of the young, the female becomes the dominant member of the pair and attacks the male savagely and noisily if he comes near, or tries to enter her nest box. She does not allow the male to enter her nest again until the young are about thirty days old and are presumably independent, and she is within eight or ten days of the birth of her next litter.

It seems, from the observations, that the female comes into heat soon after the birth of the young, and then, after mating, keeps the male away. The next litter is born about five weeks later—a long gestation period for a rat.

In Tasmania the breeding season is from October to March and each female produces more than one litter during the season.

The Broad-toothed Rat is a compact, fluffy animal with a very short tail, much like a vole of the Northern Hemisphere.

48   BROAD-TOOTHED RAT, *Mastacomys fuscus*   Group 41

The Species   Group 41

BROAD-TOOTHED RAT, *Mastacomys fuscus* (Plate 48). Central-western, western and south-western Tas., Vic., south-eastern N.S.W., (apparently confined to small local communities); mainland—under alpine conditions—heath communities of shrubs and grasses; at lower altitudes—wet sclerophyll forest with dense undergrowth containing ferns, shrubs, and grasses; Tasmania—boggy areas of sedges, button grass, tussock grass, or sphagnum moss. *Recognition:* size of rat, very compact, dark brown, fluffy, very short tail. Females with 0 pectoral and 2 inguinal pairs of mammae.

Group 42 Native-mice   Plate 49

The rest of the 'Old Endemics' (see p. 134) are much less obviously specialized than the tree-rats, the rock-rats, the hopping-mice, the stick-nest rats and the Broad-toothed Rat. They are generally soft-furred little animals, often with rather large ears; some are like big fluffy rats, others are as small as very tiny mice. No clear-cut characters separate them into groups of species and we treat them here as a single genus *Pseudomys*—a name which means 'false mouse'.

These relatively unspecialized-looking Australian native-mice are little known; details of behaviour and biology are known for few of them but, in recent years, apparently thriving populations of many of the species have been located and there is hope that this lack of information is soon to be remedied. Some species, and in particular those which formerly occurred in districts which are developed for introduced crops, or are well settled, seem to have changed in status and, like the marsupials of those places, seem to have suffered much from the works of Man—but we know little of their former abundance and we may yet take comfort from the fact that they are shy and nocturnal, and populations of them may yet be close to settlement.

The species which have been located in the wild are mostly species of the inland, of the off-shore islands and of the, as yet undeveloped, western coastal sand heaths. Thus, to mention but a few of the most important populations, the Little Native-mouse (*Ps. delicatulus*) is common at Kalumburu in the Kimberley and on the Cobourg Peninsula in the Northern Territory; Forrest's Mouse (*Ps. forresti*) is plentiful at Thevenard Island, near Onslow, W.A., and occurs near Birdsville in south-eastern Queensland and Oodnadatta in northern South Australia; the Sandy Inland Mouse (*Ps. hermannsburgensis*) is widespread and common in the Pilbara and Centre where it lives in shallow burrows often on stony ridges beneath piles of small pebbles; the Shark Bay Mouse (*Ps. praeconis*) is secure and fairly common among the coastal *Spinifex* of Bernier and Dorre islands, W.A.; the Western Chestnut Native-mouse (*Ps. nanus*) is common on Barrow Island and around Kununurra in the Kimberley; and the Ashy-grey Mouse (*Ps. albocinereus*) is numerous in a number of western coastal localities from Esperance to Shark Bay including Bernier and Dorre islands. The Tasmanian Mouse (*Ps. higginsi*) is common in the vicinity of rain forest in western Tasmania, ranging in altitude from near sea-level to about 3,000 feet.

49   FORREST'S MOUSE, *Pseudomys forresti*   Group 42

The Species   Group 42

Females with 0 pectoral and 2 inguinal pairs of mammae.

FORREST'S MOUSE, *Pseudomys forresti* (Plate 49). Arid parts of Australia (from the Pilbara, including Thevenard Island, W.A., Spencer's Gulf in S.A., to central Qld, and the Barkly Tableland, N.T.).
Recognition: size of mouse; guard-hairs slightly spiny; yellowish olive-brown above, white below; tail shorter than head and body.

LITTLE NATIVE-MOUSE, *Pseudomys delicatulus*. Northern Australia (Kimberley, northern N.T., north Qld, eastern Qld south to Rockhampton); sandy well-drained soil with sparse herbaceous vegetation.
Recognition: size of small mouse; greyish-fawn above, abruptly changing to white below; tail about same length, or longer than, head and body.

PEBBLE MOUND MOUSE, SANDY INLAND MOUSE, *Pseudomys hermannsburgensis*. Arid Australia (Pilbara and south-eastern W.A., through S.A. and southern N.T. into north-western Vic., far-western N.S.W. and south-western Qld); grasslands (including spinifex—*Triodia*), stony rises.
Recognition: size of mouse; rich yellowish-brown above, almost golden on lower sides, abruptly changing to white below; tail generally markedly longer than head and body.

NEW HOLLAND MOUSE, *Pseudomys novaehollandiae*. Eastern N.S.W. (Upper Hunter River), Port Stephens and Ku-ring-gai Chase.
Recognition: size of mouse; grey-brown above, greyish below; close short hair; white slender feet; tail length roughly one-tenth longer than head and body.

ASHY-GREY MOUSE, GREY MOUSE, *Pseudomys albocinereus*. South-west of W.A., islands in Shark Bay, W.A., south-eastern S.A., western Vic., central N.S.W., southern Qld; shrub and tree heaths, dunes.
Recognition: size of mouse; grey with lighter under-parts; soft fur; feet very pink in life, white in death; tail a little longer than head and body; white haired but pink in life, occasionally dark pigmented patches on upper surface towards body.

SMOKEY MOUSE, *Pseudomys fumeus*. Vic. (Cape Otway Ranges and Grampians); in dense sclerophyll forest.
Recognition: size of mouse; smokey-grey, greyish-white below; tail obviously longer than head and body.

WESTERN MOUSE, *Pseudomys occidentalis*. South-west of W.A.; shrub and tree heaths.
Recognition: size of large mouse; dusky grey above, paler below; tail very much longer than head and body, dark along upper-surface except near tip.

SHARK BAY MOUSE, *Pseudomys praeconis*. Peron Peninsula and islands of Shark Bay, W.A.; coastal dunes.
Recognition: size of very large mouse; brownish above, white below; short blunt head; long soft hair; tail length about equal to head and body.

GOULD'S NATIVE-MOUSE, *Pseudomys gouldii*. Southern W.A., S.A., western N.S.W.
Recognition: size of large mouse; ashy-brown washed black above, greyish-white below; tail shorter than head and body.

EASTERN MOUSE, PLAINS RAT, RIVER RAT, *Pseudomys australis*. Central and southern Qld, inland N.S.W., S.A.
Recognition: size of small rat; brownish, paler near-whitish below; ears very long (about 1 inch); tail slightly shorter than head and body.

ALICE SPRINGS MOUSE, *Pseudomys fieldi*. N.T. (vicinity of Alice Springs).
Recognition: size of large mouse; sandy-brown above, white below; fur soft and fluffy; tail considerably longer than head and body.

LONG-TAILED RAT, TASMANIAN MOUSE, *Pseudomys higginsi*. Tas.; in or near rain forest, particularly rain forests in western Tas.
Recognition: size of small rat; grey-brown, paler greyish below; tail white on under-surface, longer than head and body.

HASTINGS RIVER MOUSE, COAST RAT, FOREST RAT, *Pseudomys oralis*. North coastal N.S.W.
Recognition: size of large rat; yellowish-brown with olive tinge above, fawny-grey below; soft-furred; tail shorter than head and body.

SHORTRIDGE'S NATIVE MOUSE, BLUNT-FACED RAT, *Pseudomys shortridgei*. South-west of W.A., western Vic. (Grampians and Portland districts); swampy country surrounded by thick bush (W.A.), shrub and tree heaths of various densities (Vic.).
Recognition: size of small rat; brown with paler underside; long guard-hairs giving fluffy appearance; blunt head; tail shorter than head and body, under-surface white markedly contrasting with upper-surface; general appearance superficially like a small Southern Bush-rat.

BROWN DESERT MOUSE, *Pseudomys desertor*. W.A. (Canning Stock Route and Bernier Island), southern N.T., northern S.A., vicinity of Murray and Darling Rivers.
Recognition: size of mouse; brown with dusky wash, greyish-buff below; tail about same length as head and body.

EASTERN CHESTNUT NATIVE-MOUSE, *Pseudomys gracilicaudatus*. Central and southern Qld.
Recognition: size of large mouse; brown, paler below; slightly spiny guard-hairs which lie flat giving sleek appearance; tail length about the same as head and body.

WESTERN CHESTNUT NATIVE-MOUSE, *Pseudomys nanus**. W.A. north of Moore R. (New Norcia, Barrow Island, Kimberley), northern N.T.; among rocky hills.
Recognition: not distinguishable from *gracilicaudatus* on external features.

RECOGNITION OF INTRODUCED MOUSE OR HOUSE MOUSE

Mus musculus
Recognition: upper incisors notched behind tip by tip of lower incisor. Females with 3 pectoral and 2 inguinal pairs of mammae.

Group 43 Mosaic-tailed rats

Plate 50

The mosaic-tailed rats are a group of rat-like species whose main distribution centre is among the islands to the north of Australia; they have a coastal distribution in northern and north-eastern Australia. They are all characterized by the mosaic-like pattern of scales which they have on their tails, instead of having a regular arrangement of rings of scales as is found in other Australian rats and mice.

There are two species (*Melomys*) which are quite small, being only about nine inches from nose-tip to tail-tip, while the other mosaic-tailed species, the Giant White-tailed Rat (*Uromys*), is as large as the Water Rat.

---

*It is very likely that *nanus* and *gracilicaudatus* are a single species. However, a continuous population is not known from west to east in northern Australia and there are some recognizable skull differences between western and eastern forms. Pending further work they are left separate.

All mosaic-tailed rats climb but they are not necessarily arboreal. The Fawn-footed Melomys is said to nest in pandanus, but the Little Melomys makes nests of grass and leaves in tall grass.

Mosaic-tailed rats are vegetarian* but, as most rats will, they may also take insects and other animal foods when they are available. Their feeding habits make them important economically because of the damage they do to sugar cane and to coconuts. The Giant White-tailed Rat, of north-eastern Queensland, also raids chicken runs for eggs.

The Species   Group 43

Females with 0 pectoral and 2 inguinal pairs of mammae.

GIANT WHITE-TAILED RAT, *Uromys caudimaculatus.* North-eastern Qld (probably as far south as Townsville), Hinchinbrook Island; rain forest, venturing out into open forest.
Recognition: very large; rat-like in appearance; mosaic-like scales on tail, tail white-tipped almost hairless, often blotched white.

FAWN-FOOTED MELOMYS, *Melomys cervinipes.* North-eastern N.S.W., coastal Qld; woodland, rain forest, and trees fringing swamps etc. (e.g. pandanus).
Recognition: size of rat; greyish-brown, dense, soft, almost woolly fur; mosaic-like scales on tail; short rounded ears.

GRASSLAND MELOMYS, *Melomys littoralis.* North-eastern Qld (as far south as Rockhampton district); grasslands and grassy clearings (tall grasses).
Recognition: smaller than *cervinipes*, otherwise externally similar.

LITTLE MELOMYS, *Melomys lutillus.* Cape York Peninsula; open brushy or grassy country.
Recognition: smaller than *cervinipes*, otherwise externally similar.

POPULATIONS OF UNCERTAIN STATUS AMONG *Melomys:* populations of *Melomys* occur in the Northern Territory (including Groote Eylandt and Melville Island) and in the Kimberley of W.A. These have received various names such as *Melomys burtoni, Melomys melicus, Melomys cervinipes albiventer, Melomys mixtus.* Further research is necessary before the identity of these can be established, but there seems to be little doubt that they will be found to belong with *Melomys littoralis* or *Melomys cervinipes.* Mosaic-tailed rats of a Northern Territory population are shown in Plate 50.

---

* Trying to decide the 'normal' food of wild mammals is sometimes quite a problem because many mammals are opportunistic in this regard. Professor John Harrison has discussed the difficulties of interpreting stomach contents in establishing the usual diet of various species. He has concluded that the earth he found in the stomachs of bandicoots and *Rattus sordidus* was, presumably, swallowed accidentally and should not be regarded as food. He concluded that, similarly, those animals which appear, from their stomach contents, to take small quantities of insect food may well do so accidentally; much as weevils are eaten with porridge, although most men would indignantly deny that they were insectivorous. But the eating of other rats by *Rattus sordidus* which otherwise appears to be strictly vegetarian, can, however, hardly be regarded as accidental. He asked whether it could not be reasonable to consider the eating of a defeated foe as a natural extension of a fight conducted with teeth (and not a meal), in the same way as the eating of young rats by the mother is an extension of eating the placenta, and not cannibalism!

10

Australian bats

HUGE POPULATIONS of small night-flying bats haunt the air over the Australian countryside from twilight to the early morning. But it is only when a bat takes shelter in a house by mistake, or a population takes up residence in a church, or country hall, that the majority of people become aware that they are there. Yet you have only to stand by a country pool on a still warm evening, to see for yourself something of their numbers.

For many years, Museum taxonomists in Australia and overseas have worked to determine the species of Australian bats, but it is only in recent years that a number of enthusiasts, particularly in the Division of Wildlife Research, C.S.I.R.O., the universities, and among the various amateur speleological groups, have begun to study populations in the field, to elicit their breeding behaviour, and by banding and releasing and recovering them, to learn their movements.

Bats are the only mammals which truly fly. They differ from the gliders in that they do not merely depend upon gravity, and the initial leap, for movement through the air but fly by supporting and propelling themselves with flight surfaces rather like birds. Unlike birds these flight surfaces are not only modified fore limbs but consist of a thin membrane (see Plate 52, p. 163) which is an extension of their body skin all the way along both sides of the body from the shoulders backwards. It is most extensive from behind the tips of the first fingers to the ankles and the fingers are greatly elongated and are held out to stiffen it like the ribs of an umbrella. The hind legs and tail are often involved in the membrane as well; in some species it forms a convenient pocket between the legs—which may be used to hold prey even during flight. Those bats which have a tail membrane usually have it stiffened by two slender rods of bone, peculiar to bats, called the calcar; these stick out towards the tail from each heel. The degree to which the calcar is developed, and the amount of the tail which is incorporated in the membrane, are features useful in bat identification. The flight membranes of bats are delicate and wonderful structures; they consist of two layers of skin which are extensions of the upper and lower surfaces of the body. These skins look like a single membrane through which run the delicate blood vessels which provide them with nourishment.

In Europe, and in America, much has been learned of the biology of bats, and from this work fascinating accounts have emerged of echo-locating systems which depend upon the ability of bats to emit high-frequency sounds, and to interpret their echoes as they bounce back from obstructions. Much is also known about the way in which bats breed, and even control their temperatures in their secluded roosts; we can now hope that, before many years have passed, similar knowledge will be obtained for many of the Australian species.

Most species of bats are very small and are insectivorous. These are called Microchiroptera (meaning 'little hand-winged' animals), and when people talk about bats in belfrys, these little bats are what they mean. But there are also other bats—fruit bats—which in Australia, are more usually called flying foxes or blossom bats. These Megachiroptera ('large hand-winged' animals) are specialized feeders upon fruit and blossoms. In general, flying foxes and their relatives roost by hanging in trees while the microchiropterans hide in holes. Various species exhibit strong habitat preferences; thus some are tree bats, usually being found in holes or under bark, while others are cave bats and are usually found in caves or mine shafts.

The species of Microchiroptera have wings which span generally less than one foot; Megachiroptera are almost all bigger than this and their wings may span up to four feet. The Australian exceptions to this general rule are the Ghost Bat, a large microchiropteran (p. 160), and small megachiropteran blossom-eating bats (p. 180).

The ears and faces of microchiropteran bats are often strangely elaborate; for example, there are complex folds of skin around the nostrils of some (e.g. horseshoe bats and the Ghost Bat) and it is now known that in the horseshoe bats, who emit their ultrasonic echo-locating pulses through the nose, these are used to concentrate the bursts of energy in a narrow beam. Presumably the complex projections, such as the tragus and other folds, which are parts of the ears of some bats, have a role in receiving echoes. The tragus is the lobe of the ear which lies in front of the ear passage and, in man, partly covers it; in bats it is often long and of characteristic shape. Like the calcar and the tail it is useful as an identifying feature.

Bats have a long history; their earliest fossils are from the Eocene period and are over fifty million years old. Already at this early time of mammal history, bats were fully equipped for specialized flight. From the structure of their teeth it seems likely that they evolved from the Insectivora, an Order of mammals to which shrews, hedgehogs and true moles belong. None of these relatives of bats occurs in Australia, and their absence from our fauna contributes to our current understanding that the bats evolved outside Australia and flew in as colonists.

Nowadays a great deal is said about the advantages to man of controlling noxious insects by biological means instead of through pesticides; and birds, and various insects such as ladybirds, are cited as important agents; but little is said of the role of bats. Yet bats are voracious feeders which research overseas has shown are capable of eating one-third of their own bodies' weight of insects in an hour. At even less than this rate a colony of ten thousand Bent-wing Bats cannot fail to have a significant effect upon insect populations flying in the vicinity.

In our inherited European traditions bats are nasty things; unnatural, silent, flittering creatures of the night. By reputation, too, they are verminous and even blood-suckers. Try to forget these fears which have come out of ignorance. If you get a chance to examine a bat closely, look and see it as it really is—a very wonderful little mammal which is different because it has conquered the air.

We still have a long way to go in the knowledge of the distribution of Australian bats and even of the arrangement of their species. In respect of the latter, no two authorities on Australian bats even agree on the numbers of species in the genera. In preparing the lists of species which follow, and in recording their habitats, I am grateful for the advice which Mr J. McKean of the Division of Wildlife Research, C.S.I.R.O., has been able to give me from his very great personal field experience of the eastern and northern bats; and also for the help which Mrs John Bannister has given me with the classification of some of the genera.

Group 44 The Ghost Bat   Plate 51

The Ghost Bat is the only Australian member of a family (Megadermatidae) which occurs otherwise in Asia and Africa. It is a very large bat for a microchiropteran, most of which have a wing span of well under ten inches; adult Ghost Bats have a wing span of about two feet. Ghost Bats are highly specialized carnivores; they prey upon lizards and small mammals on the ground and, at night, on birds as they roost and even on other bats.

Like most carnivorous mammals, the Ghost Bat is extremely wary and is very difficult to approach in the wild; it does not hesitate to fly into the daylight away from an intruder into its cavern.

In the Pilbara district of Western Australia, the Ghost Bat is of the pale 'ghostly' desert form with ashy-grey back and white under-parts. Juveniles are sooty-grey all over. In the northern part of its range, and in particular in the Kimberley, in Arnhem Land, and around Rockhampton in Queensland, the adult Ghost Bats are as dark as, or even darker than, juveniles from the desert.

Field observations on the Ghost Bat in the arid Pilbara, and in the Kimberley of Western Australia, by Mr Athol Douglas of the Western Australian Museum, showed that females give birth to a single young about November in southern areas, and about September in the far north; it seems likely that at these times the females congregate without the males in maternity colonies. While they are still suckling, the young are left in the roost while the females go out to hunt. Later, during weaning, when the young start feeding upon a meat diet, the females return to the roosts with their prey for the young to eat. Mr Douglas has recently found that, in the Pilbara, by January, the young, which are by then nearly as large as their mothers but dark grey in colour, are hunting in couples with their mothers. At this stage they are still living in large colonies—apparently without adult males. Douglas also found large all-male colonies at the same time of the year.

The Ghost Bats emerge from their shelters after sunset, singly or in small groups, and return before dawn. They drop on mammals from above, enveloping them with their flight membranes, and kill them with bites about the head and neck. They then take the victim to a high point, or back to the roost while they are rearing young, and feed there—the bits that fall from the meal are ignored and add to the accumulation of dung, feathers, bones and other debris, which are so characteristic of the caves frequented by female and young Ghost Bats. They eat large amounts of food; everything, including flesh, bones, teeth, fur, small feathers and the chitinous exoskeletons of insects, is swallowed and passed through the gut. Ghost Bats appear to need this roughage in their diet because if they are fed on boneless meat in captivity they soon become distressed and fouled with loose excreta. The species of birds most frequently killed have been shown by Douglas to be the Budgerigar and the Owlet-Nightjar— even the relatively large Red-plumed Pigeon is taken.

The Ghost Bat was formerly believed to be a very rare species (see p. 25) but it is now known to occupy a large area of arid and tropical Australia. Long before European times, it occurred in localities as far south as the South West of Western Australia and the Flinders Range in South Australia; it is not understood why the range has

51   GHOST BAT, *Macroderma gigas*   Group 44

has suggested that although conditions may still be climatically suitable in the southern areas, changes in vegetation may have made it more difficult for the Ghost Bat (and also night-haunting birds of prey like owls) to capture prey in denser vegetation. Such vegetational changes could easily have resulted from an alteration in the frequency with which the forests were burned, following the arrival of aboriginal man in Australia and his use of fire for hunting.

Ghost Bats are secretive and, being exclusively cave-haunting, are so unobtrusive that residents, even miners who enter tunnels in which they occur, will claim to know of the presence of other bats but not of Ghost Bats. Colonies occur principally in caves, deep fissures in rocky walls, and in mine shafts and tunnels. On rare occasions, they will occupy buildings as feeding stations but not as roosts; thus, at Kalumburu, the missionaries of the Catholic Mission were kept awake at night by Ghost Bats noisily eating prey in the Church entrance, and in the corridors, and dropping the remains of their prey on the floor. Douglas recounts that these church-haunting episodes were confined to the wet (summer) season and it seems likely that there is some seasonal movement of colonies or, at least, a change of hunting behaviour between summer and winter.

The Species   Group 44

GHOST BAT, FALSE VAMPIRE, *Macroderma gigas* (Plate 51). W.A., N.T., Qld—all north of 29°S.; caves and mineshafts.
Recognition: ears joined together above forehead; no tail; nose-leaf.

Group 45 Long-eared bats   Plate 52

The Australian long-eared bats belong to a sub-family (the Nyctophilinae) which, apart from a single North American genus, only occurs in Australia and New Guinea. Australian long-eared bats can easily be distinguished from all other Australian bats except the Ghost Bat (and that is easily told by its size and colour) and the Northern Mastiff Bat by the fact that the ears are joined together by a membrane above the level of the forehead. The best way to test whether a bat has joined ears is to slide a blunt-pointed instrument, such as a pencil, forward along the crown of the head so as to pass between the ears; in long-eared bats this is obstructed and is unable to pass through on to the muzzle.

In southern Australia there are two very distinct species, one larger (the Greater Long-eared Bat), and one smaller (the Lesser Long-eared Bat). In addition, there are two (and possibly three) other species in North Queensland and the Northern Territory.

The long-eared bats are mostly tree bats, but both Greater and Lesser Long-eared Bats have been found in caves in Western Australia and the Lesser Long-eared Bat even under a stone on the ground, in open country. In the South West of Western Australia, the Lesser Long-eared Bat is the species of bat most commonly found in

52 GREATER LONG-EARED BAT, *Nyctophilus timoriensis*

homes where it takes refuge in curtains, or in coats hanging on verandas.

In Tasmania the Lesser Long-eared Bat is, next to the Little Bat, the most common species of microchiropteran. Recent studies by Mr R. H. Green of the Queen Victoria Museum, Launceston, have shown that it forms maternity colonies in early summer which contain both females and males; the colonies examined all occurred within buildings. When these colonies were disturbed the females removed their young to alternative sites. Females, with the young holding to their under-surfaces, fly with ease; the young cling on with claws and milk-teeth which have sharply recurved cusps.

Females usually give birth to twins which are naked at birth. They are born tail foremost into the tail membrane of the mother which she hooks forward beneath her. It appears that the newly-born bat frees its own head from the mother, after her contractions have ceased, and it then crawls, still attached by the umbilical cord, forwards along the under-surface of the mother to locate and seize a nipple which is on the chest and well to the side under the armpit. The umbilical cord soon becomes bloodless and then parts.

Nothing is known of the behaviour of the Lesser Long-eared Bat at other times of the year and no information is yet available on its movements; between 1957 and 1963 thirty-two Lesser Long-eared Bats, and one Greater Long-eared Bat, were banded under the Australian Bat-banding Scheme but only three of these specimens have been recovered; the most notable of them was a Lesser Long-eared Bat which was caught at the same locality ten months after it was banded.

When they are hunting, lesser long-eared bats will land on the ground to pick up beetles and other insects. They spring into the air with their wings folded, jerking themselves up from their wrists—then when they are in the air they open their wings.

Although long-eared bats have no clearly marked nose-leaf like the horseshoe bats or the Ghost Bat, they have small rudimentary projections on the upper surface of the snout.

The Species   Group 45

GREATER LONG-EARED BAT, *Nyctophilus timoriensis* (Plate 52). South West of W.A., S.A., Vic., Tas., eastern N.S.W., eastern Qld north to Innisfail; hollow trees.
Recognition: ears joined together above forehead; forearm over 1¼ inches; nose-leaf a well-marked ridge behind nostrils and then posterior part behind that little more than a bump.

LESSER LONG-EARED BAT, *Nyctophilus geoffroyi.* The whole of southern and central Australia including N.T. and north-western Qld to the latitude of Normanton at the Head of the Gulf of Carpentaria; trees, occasionally in caves, crevices in rocks.
Recognition: as in *timoriensis* but size smaller, forearm mostly under 1½ inches; small Y-shaped posterior nose-leaf quite well developed, forward part a ridge as in *timoriensis*.

NORTH QUEENSLAND LONG-EARED BAT, *Nyctophilus bifax.* Northern Qld from Atherton near Cairns to Darwin, N.T.; presumably trees, found in houses.
Recognition: not separable externally from *timoriensis* except by slightly smaller size.

ARNHEM LAND LONG-EARED BAT, *Nyctophilus arnhemensis.* Northern N.T. (Arnhem Land and Groote Eylandt); under loose bark of paper-bark trees (*Melaleuca*).
Recognition: ears joined; size small like *geoffroyi* but said to have rather simpler nose-leaf and shorter ears.

NAME OF UNKNOWN STATUS *Nyctophilus walkeri:* the status of the specimen from Adelaide River in the Northern Territory which was described as a separate species *Nyctophilus walkeri*, is unknown. It needs re-examining in the light of what we now know of the populations of the Arnhem Land and North Queensland Long-eared Bats.
Recognition: ears joined; size very small, forearm less than 1½ inches; not separable externally from *arnhemensis.*

Group 46 Horseshoe bats

Plate 53

Numbers of bats are characterized by curious projections of skin which grow outwards from the snout around the nostrils. This organ is called the nose-leaf; its function is not fully understood but it is suspected that structures like it, and the complex ear structures often found in bats, play an important part in the echo-location by which bats not only find their way around obstacles, but also locate and capture their prey. In Australia, four groups of bats have nose-leaves but none are more prominently developed than in the two families Rhinolophidae and Hipposideridae which together are called horseshoe bats; the other bats with nose-leaves, namely the Ghost Bat and the long-eared bats, have already been mentioned. In the long-eared bats, the nose-leaf is so small that it is little more than a projection on the top of the nose. The name Horseshoe refers to the semicircular shape of the lower part of the nose-leaf.

Horseshoe bats are often brightly coloured. They are mostly greyish-brown but some species have a bright rufous or even orange colour phase. For example there are both rufous and grey phases in the Eastern Horseshoe Bat, while the Dusky Horseshoe Bat ranges from a sombre brown to a bright orange. In the Orange Horseshoe Bat, the orange form predominates over a less common fawn one. It is not clear whether the orange and grey-brown forms represent distinct colour varieties of the bats or whether they are stages of moult. In recent work in Queensland, Messrs J. L. McKean and W. J. Price collected and examined numbers of the Eastern Horseshoe Bat; they found adults of both colour phases while all yearlings were grey-brown and some of these were moulting into the orange phase.

The complex nose-leaves of the horseshoe bats are made up of three main parts. There is a lower part which is large and horseshoe-shaped (it may also have additional little leaves under its sides); there is a central part around the nostrils which may carry a projecting organ; there is an upper (or hindmost part) on the forehead which sometimes also has an upward projecting piece in the centre. The detailed arrangement of these parts and their variations are useful in identification.

No Australian horseshoe bats have ears which are joined together above the forehead like those of the long-eared bats.

Male Dusky and Fawn Horseshoe Bats have a peculiar sac on the head behind the hindmost portion of the nose-leaf. It can be extended or lowered at will by the bat and can exude a clear, apparently odourless, liquid. Its function is quite unknown. Females of these species have a slight depression filled with hairs in that position.

The arrangement of species which I adopt here is that proposed by Mr J. E. Hill of the British Museum (Natural History).

The Species   Group 46

SECTION 1: Rhinolophidae

EASTERN HORSESHOE BAT, *Rhinolophus megaphyllus*. Eastern Australia (from Vic. to Cape York—all but a few records are to the east of the Great Dividing Range); mines, caves or tunnels.
Recognition: ears not joined together; nose-leaf in three parts, the lower horseshoe-shaped, central part above nostrils has a forward projecting process, the upper part pointed sticking up from forehead; forearm under 2 inches.

LARGE-EARED HORSESHOE BAT, *Rhinolophus philippinensis*. North-eastern Qld, (north of Townsville); mine tunnels, probably caves.
Recognition: differs from *megaphyllus* in having upper part of nose-leaf rounded; larger ears; forearm over 2 inches.

SECTION 2: Hipposideridae

DUSKY HORSESHOE BAT, *Hipposideros ater*. Qld (Cape York Peninsula north of Innisfail) N.T. (northern), W.A. (Kimberley).
Recognition: ears large, rounded at tips, furred only at base, not joined together; coat colour pale grey with darker tips, or orange; nose-leaf large, squarish in outline, no forward projections, no supplementary leaflets at sides of lower part.

FAWN HORSESHOE BAT, *Hipposideros galeritus*. Cape York Peninsula (north of Cooktown) caves, mines and deserted dwellings.
Recognition: ears broad, sharply pointed, hairy for three-quarters of their length, not joined together; nose-leaf with 2 pairs of supplementary leaflets at sides of lower part.

LARGE HORSESHOE BAT, DIADEM BAT, *Hipposideros diadema*. North-eastern Qld (as far south as Cardwell); mines.
Recognition: ears pointed, not joined together; nose-leaf with 3 or 4 pairs of supplementary leaflets at sides of lower part.

WARTY-NOSED HORSESHOE BAT, *Hipposideros semoni*. Northern Qld (Townsville north-wards); rain forest, tree dwelling, deserted buildings.
Recognition: ears not joined together, very long and narrow with acute point, haired for only one-third of length; nose-leaf with 2 pairs of supplementary leaflets at sides of lower part, central part above nostrils has well-developed club-shaped forward projection, posterior part on forehead with well-developed club-shaped projection from centre of its upper edge.

LESSER WARTY-NOSED HORSESHOE BAT, *Hipposideros stenotis*. Kimberley of W.A. (Derby), N.T. (Arnhem Land), north-western Qld (Mount Isa); caves or rock-shelters.
Recognition: scarcely separable from *semoni* except for smaller size and very small size of uppermost club-shaped projection.

ORANGE HORSESHOE BAT, *Rhinonicteris aurantius* (Plate 53). W.A. (northern Kimberley), N.T., (? South Australia*); caves during wet season.
Recognition: bright orange or brown; ears not joined above forehead; nose-leaf between eyes much divided into complex parts (see illustration), central part above nostrils has forward projecting portion.

*See footnote to p. 148.

53   ORANGE HORSESHOE BAT, *Rhinonicteris aurantius*   Group 46

Group 47 Mastiff bats, scurrying bats Plate 54

The mastiff bats are so-called because some species have very heavily wrinkled lips and square muzzles rather like those of Mastiff dogs. They are unusual bats because some species appear to be very much at home on the ground. The White-striped Bat, for example, scurries around almost like a rodent. The movement of its forearms is rather deliberate like a skier walking on rough ground. To take off, these big bats spread their wings out along the ground and become air-borne with a single pro-pulsive stroke.

Although some mastiff bats do not have heavily wrinkled lips, all can be recognized very easily by the projecting thick, almost rodent-like, tail. This sticks out well behind the tail membrane which is tucked up rather closely round its base; it does not project through the membrane like that of the sheath-tailed bats (see p. 170), but out beyond it. The largest of the Australian mastiff bats, the White-striped Bat, looks extremely fierce, but it is one of the gentlest bats that I know; although I have handled many of them, none has ever attempted to bite. This bat is called the White-striped Bat because, on either side of the body under the wings, there is almost always a narrow band of pure white fur which contrasts vividly with the rest of the bat which is black or deep velvety brown. This band of fur can only be seen when the wings are fully spread.

The Species Group 47

The arrangement of the species followed below is that proposed by Dr Heinz Felten of the Senckenberg Museum, Germany.

WHITE-STRIPED BAT, *Tadarida australis* (Plate 54). The southern half of W.A., S.A., southern Qld, N.S.W., Vic.; tree hollows.
Recognition: usually a white stripe of fur on either side of body under wings; fur velvety black or brown; ears separate above forehead; large, forearm over 2 inches; gular pouch.

NORTHERN MASTIFF BAT, *Tadarida jobensis*. Northern Australia (northern W.A., N.T., northern Qld—possibly including northern S.A.); tree hollows.
Recognition: brown all over; ears joined together above forehead; size a little smaller than *australis*; no gular pouch.

LITTLE FLAT BAT, *Tadarida planiceps*. Southern W.A., S.A., southern N.T., N.S.W., Vic.; hollows and crevices in trees.
Recognition: ears pointed, separate; small, forearm less than 1½ inches; whole body very flat; no gular pouch in males or females.

LITTLE NORTHERN SCURRYING BAT, LITTLE NEW GUINEA SCURRYING BAT, *Tadarida loriae*. Northern Australia (Kimberley, W.A., N.T., northern Qld); roofs of houses, possibly dead trees.
Recognition: as *planiceps* but gular pouch sometimes in males.

NORFOLK ISLAND SCURRYING BAT, *Tadarida norfolkensis*. Norfolk Island (but there is some doubt as to the correctness of this locality).
Recognition: as *planiceps* but gular pouch in only known specimen which is male.

54   WHITE-STRIPED BAT, *Tadarida australis*   Group 47

Group 48 Sheath-tailed bats   Plate 55

The sheath-tailed bats are very easy to recognize by two characteristic and conspicuous features. First of all, they have long dog-like noses rather reminiscent of
the flying foxes and, secondly, they have a tail which although superficially free, like
that of the mastiff bats, is actually in a sheath which penetrates *through* the tail
membrane and extends from its upper surface. If you find a bat with a projecting tail
and wish to see which sort it is, run a finger along the under-surface of the tail, away
from the body. If the animal is a sheath-tailed bat, the tail membrane will be found
to form a pouch lying free underneath the tail, instead of being joined to it as it is
in other bats.

One of the sheath-tailed bats, the Common Sheath-tailed Bat, of the caves of
central and northern Australia, is very familiar to all who enter caves and mines.
It is a very alert bat and when disturbed by an intruder rocks up and down on its
forearms and runs around on the surface of the rocky walls, with extraordinary
rapidity, on its wrists and its hind feet.

The Yellow-bellied Sheath-tailed Bat is one of the most handsome of our bats and
also one of the least known. It is probably a tree bat.

All species have glandular, gular (or throat) pouches in one or both sexes.

The Species   Group 48

COMMON SHEATH-TAILED BAT, SHARP-NOSED BAT, *Taphozous georgianus* (Plate 55). W.A.
(Murchison District and north), N.T., Qld; caves and mines.
Recognition: brown all over; gular pouch present in some males.

YELLOW-BELLIED OR WHITE-BELLIED SHEATH-TAILED BAT, *Taphozus flaviventris*. Throughout
Australia excluding Tasmania; hollow trees.
Recognition: white or yellow belly contrasting vividly with black back; prominent gular
pouch.

NORTH-EASTERN SHEATH-TAILED BAT, *Taphozous australis*. North-eastern Qld (south to
Cardwell) and Torres Strait Islands; caves.
Recognition: brown all over; gular pouch present in males.

NAKED-RUMPED SHEATH-TAILED BAT, *Taphozous nudicluniatus*. North-eastern Qld (from
Townsville northwards); edge of rain forest.
Recognition: reddish-brown mottled with white patches; rump almost naked; male with
gular pouch.

TROUGHTON'S NEW GUINEA SHEATH-TAILED BAT, *Taphozous mixtus*. Cape York Peninsula;
flying in vicinity of 'open forest near a gully filled with fringe forest' (note of collector).
Recognition: brown above, belly paler but contrast not as marked as in *flaviventris*;
gular pouch.

Group 49 Ordinary small bats   Plate 56

The bats in this group all belong to the family Vespertilionidae. This family also
includes the long-eared bats, but, because of their characteristic appearance, it is

55 COMMON SHEATH-TAILED BAT, *Taphozous georgianus*   Group 48

easier to consider those separately (as the sub-family Nyctophilinae—see p. 162). Vespertilionid bats, which we are calling ordinary small bats here, are all rather unspectacular in appearance; they have generally short ears, short, blunt and some- times very broad noses, and tails enclosed within their tail membranes. Some can easily be identified by size (quite large, or very small), others by colour, and yet others by the structure of their wings. Identifying these little bats is not an easy job and, in the end, it almost always requires the skull of the specimen to be examined in order to confirm the identification.

Some of the little bats are exceedingly common; for example, the Bent-wing Bat *Miniopterus schreibersii* is one of the most common cave bats in the eastern coastal region. It is the species which has been most studied by biologists in eastern Australia. From these studies Dr P. D. Dwyer, when he was at the University of New England, has shown that the eastern population of this species depends upon a comparatively small number of maternity colonies, each of which serves a large area. It is obvious that this kind of specialized breeding behaviour makes whole sections of the popu- lation very vulnerable and it is urgent that extensive searches should be made to discover the distribution of maternity colonies in order to have them protected.

Another of the most common and widespread species of ordinary small bats, the Little Bat, has recently been studied near Kelso in northern Tasmania by Mr R. H. Green of the Queen Victoria Museum, Launceston. He discovered that the bat forms congregations in summer in the cavities of walls, holes in trees, and similar places. These too are maternity colonies and, as with the Bent-wing Bats, the females gather in them to give birth to their young.

The young are born towards the end of November; they are naked and blind, but are agile and grip tenaciously to the nipples and fur on the underside of the mother. To increase this grip the young have specially developed milk-teeth; the incisors have three long prongs which are curved backwards like miniature bag-hooks, and the lower canines are similarly curved and have two sharp prongs. Green says that the parents are quite capable of flying with young so attached but do not normally do so. Young are normally left in the roost when the adults are out hunting. The young bats make their first independent flights when they are about seven weeks old.

Mr Green does not know what happens to them after they leave the maternity colonies, but he suspects that they do not migrate, but merely disperse into such places as crevices in trees where they may even partially hibernate over the winter.

Although the Little Bat is common and widespread its colonies are small; those studied by Green contained less than fifty individuals. By contrast huge colonies of the Bent-wing Bat which have been counted have contained as many as 18,000 bats; such concentrations provide remarkable opportunities for study and, in investigating a population in north-eastern New South Wales, Dr Dwyer was able to mark 8,775 individuals to gain information on life history, breeding and seasonal movements in the species. Mating occurred in this population in early winter and the resulting embryos were very slow in their development until the spring when they began to develop at a more usual rate for a mammal. During October and November, towards the end of their pregnancy, females make their way to special maternity caves in which they congregate to give birth. The first young are born in early December. At night the young bats are left in the caves while the mothers go out to feed; they grow rapidly and they are attended and suckled until about the middle of March when the

TASMANIAN PIPISTRELLE, *Pipistrellus tasmaniensis*   Group 49

mothers desert their young and leave the maternity caves altogether. But even before they are deserted, in fact soon after birth, the adults and young segregate themselves into separate clusters within the cave when they are not actually nursing; these clusters contain enormous numbers of individuals and, by concentrating a great mass of energy in the form of living bodies, the clusters probably play an important role in maintaining the stable and high temperatures which are needed to make it possible for the young to grow. It is probable that the temperatures in the caves are raised, and kept high, by the massed metabolic activity of the bats themselves. To give some idea of the concentration involved, the clusters of young in these colonies reach a density of 368 baby bats per square foot of cave ceiling. In one cave worked on by Dr Dwyer the cluster of juveniles alone was twenty-four feet long and one and a half feet wide and was estimated to contain 11,800 baby bats. Although it seems scarcely credible, it is very probable that females succeed in locating and feeding their own young from out of this great mass. Clusters are not only characteristic of maternity colonies but are also a feature of other colonies; but at other times they are smaller and are less frequent in winter, and most frequent and largest in summer.

By the seventh week of their lives the juveniles are able to fly reasonably well and are making short foraging flights. Before this they learn to fly in the cave; but many die in these experimental flights because they fall to the floor where some become ensnared in the heaps of semi-liquid guano, and others are drowned in pools of water. Those who do not become trapped try to fly from the ground, or crawl backwards in an attempt to reach the walls where they can work their way upwards to regain the massed clusters of juveniles on the roof.

When the maternity caves are first occupied they contain some juveniles from the year before but later they contain only mothers and their young. Females do not start to breed until their second year.

Bent-wing Bats do not occupy the same roosts the year round and they have complex patterns of movement among the caves of the area in which they live. Individual bats may travel forty miles in a single night. The sexes, however, behave rather differently. Males, in particular, seem to become attached to particular sites, once a period of juvenile wandering is over, but females move widely over the total area centred about a particular maternity colony. Adult females have been known to move 200 miles to their maternity cave and one female banded at a maternity colony was recovered 370 miles away. It seems that individuals are familiar with the location of many sites within the area occupied by the population and some caves even seem to be used as regular staging camps in such major movements as those made by females towards maternity colonies.

In winter the bats tend to move into caves which become as cold as possible; in such places they become torpid by allowing their body temperatures to fall close to that of the surrounding air; the resulting low metabolic rate allows them to conserve stored reserves of fat at a time when food is scarce; they have built up these reserves in the previous summer. Caves which are most suitable for winter are those which have air moving freely through them and usually have large openings. In order to find the best conditions it is sometimes necessary for bats to move from one cave to another in different seasons; but sometimes a move to a different part of the same cave will achieve the desired result.

In summer the bats move into places which make it easy for them to maintain a high body temperature, with the least expenditure of energy, because at this time they are alert and active all day and regulate their body temperatures at about 30°C. (86°F.) at which temperature they can move freely. Domes, or deep indentations, in the roofs of caves, where the warm air, heated by their bodies, remains around them are places which provide such conditions. The bats cluster in these.

During the period of sexual activity in the autumn and early winter (April to mid-June) special mating colonies are established by males; these are visited by transient females seeking impregnation.

Despite their torpid condition in winter, and somewhat sluggish state in autumn and spring, the bats may go out to forage when conditions are suitable at any time of the year. They time their emergence from the caves so that they leave within thirty-five minutes after sunset and then return a few hours before dawn. Thus, in winter, the most efficient sites are those which provide the lowest temperatures (to allow for effective torpor and conservation of reserves) while being situated in the warmest climates (to give the most abundant insect food on the winter and spring nights that they are able to fly). Therefore it is not surprising that in north-eastern New South Wales Bent-wing Bats move from the Tableland down to the coast in winter.

Dr Dwyer has also made a comparable study of the Little Bent-wing Bat and he found that there were a number of important differences between the two species where they occur together in northern New South Wales. The smaller species is confined to the sub-tropical coastal region and does not extend up into the Tableland or on the slopes. There the Little Bent-wing does not have so pronounced a period of winter torpor and feeding persists longer into the cold season. In addition, the period of slowed-down embryonic development, which occurs in winter months, is much shorter in the Little Bent-wing than it is in the Bent-wing Bat. In fact, in the tropics there is no delay at all. Since the only known nursery colony of this bat occupies a cave jointly with the much more numerous Bent-wing Bat, the Little Bent-wing may even depend upon the much larger numbers of the Bent-wing Bat to keep the temperature at a level suitable for juvenile growth. Dr Dwyer suggests that the persistence of the Little Bent-wing at the southern limit of its range in New South Wales may be dependent upon the facilities provided by its larger and better adapted relative.

Dr Dwyer has contrasted the behaviour of these two species of bent-wing bats with the Little Pied Bat which, although a cave bat like the others, has very much smaller breeding colonies and does not generate the incubator-like atmosphere of the maternity caves of the bent-wings. In fact, one breeding colony of the Little Pied Bat recorded by him had only thirteen females, two males and twenty young in it; it formed a cluster in a small indentation in the ceiling of an abandoned diamond mine which had not only a low humidity but a temperature of 60°F. Dr Dwyer suggests that, in these differences, we can detect the origins of these two kinds of bats. The bent-wings are tropical bats which have colonized colder southern latitudes, while the Little Pied Bat is a true inhabitant of temperate climates.

The Species

Group 49

BENT-WING BAT, *Miniopterus schreibersii*. Eastern Australia east of the Great Dividing Range (from northern Qld, to southern Vic., south-eastern S.A., north-western N.T., west Kimberley); caves, mines, occasionally houses.

Recognition: last joint of 3rd finger very long (more than twice the length of second-last joint) and when wing folded, folds forward to lie between the straight 2nd digit and the forearm; forearm between 1½ and 2½ inches long.

LITTLE BENT-WING BAT, *Miniopterus australis*. North-eastern N.S.W., Qld (east of the Dividing Range); caves, mine workings, occasionally houses.
Recognition: 3rd digit as in *schreibersii*; but smaller, forearm between 1 and 1½ inches long.

GOULD'S WATTLED BAT, *Chalinolobus gouldii*. Australia-wide; tree hollows, often hanging among leaves in daylight.
Recognition: dark mantle across head and shoulders, brownish posterior part (in the north contrast between these is reduced, darker all over); prominent lobe of skin joining ear to edge of lip.

CHOCOLATE BAT, *Chalinolobus morio*. Southern Australia including Tas.; tree hollows, and in parts of W.A. and S.A., in caves as well.
Recognition: Brown and small; small and inconspicuous lobe at corner of lip; most have a ridge of fur across snout in front of eyes; ears too short to meet above head when pressed together.

LITTLE PIED BAT, *Chalinolobus picatus*. North-western N.S.W., western Qld; caves, mine tunnels.
Recognition: small, forearm less than 1¼ inches; velvety black with variable amount of white fur forming white crescent along sides under wings and tail membrane; prominent lobes or wattles at hind edge of lip; ears too short to meet above head when pressed together.

LARGE-EARED PIED BAT, *Chalinolobus dwyeri*. Central N.S.W., and adjacent part of Qld.
Recognition: colour as in *picatus* but ears very large; size larger, forearm over 1½ inches.

HOARY BAT, FROSTED BAT, *Chalinolobus rogersi*.* North Kimberley, northern N.T.; all specimens collected in flight.
Recognition: colour black or dark grey with white tips to fur—obviously frosted appearance except when wet; very small lobes or no lobes at corner of lip; small, forearm less than 1½ inches.

LITTLE BAT, *Eptesicus pumilus*. Australia wide; caves, Martins' nests, buildings, tree hollows.
Recognition: very small; fur dark, grey-based with brown tips giving over-all brown colour; ears long enough to meet over top of head when pressed; otherwise no obvious recognition characters.

TASMANIAN PIPISTRELLE, *Pipistrellus tasmaniensis* (Plate 56). South-eastern Qld, eastern N.S.W., southern Vic., south-western W.A., Tas.; hollows in trees, occasionally caves.
Recognition: dark brown; two pairs of incisors between upper canines; large, forearm about 2 inches; outer edge of ear with notch near its tip.

YELLOW-HEADED PIPISTRELLE, *Pipistrellus javanicus*. No habitat recorded (locality merely given as 'Australia', see page 204).
Recognition: not separable externally from *Eptesicus pumilus*.

TIMOR PIPISTRELLE, *Pipistrellus tenuis*. W.A. (Dampier Land); no habitat recorded with only known specimen.
Recognition: not separable externally from *Eptesicus pumilus*.

PAPUAN PIPISTRELLE, *Pipistrellus papuanus*. Cape York; roofs of houses.
Recognition: base of thumb and sole of foot have small, barely visible, cushion-like suckers; not separable externally from *Eptesicus pumilus* with certainty.

---

*Another bat called *Chalinolobus nigrogriseus* is known from eastern Qld and north-eastern N.S.W. Its status is very uncertain. Some authors place it with the Chocolate Bat, but it may represent the eastern end of the range of the Hoary Bat.

LARGE-FOOTED MYOTIS, *Myotis adversus*. Eastern S.A., Vic., western N.S.W., eastern Qld, northern N.T., (possibly W.A.); in caves, crevices (in between joists of houses).
Recognition: small, nondescript; long muzzle; forearm about 1½ inches long; heel bone which supports tail membrane very long, about ¾ length from heel to tail.

SMALL-FOOTED MYOTIS, *Myotis australis*. N.S.W.; habitat not recorded.
Recognition: said to differ from *adversus* by shorter heel bone which only extends half-way between heel and tail, wings attached to feet at base of toes instead of to ankles.

GREATER BROAD-NOSED BAT, *Nycticeius rueppellii*. Eastern N.S.W., south-eastern Qld; hollows of trees, one of the first bats to be seen after sundown.
Recognition: large, forearm about 2 inches long; brown; only 1 pair of small incisors between the upper canines as in all *Nycticeius*; muzzle wide and thick at sides.

LITTLE BROAD-NOSED BAT, *Nycticeius greyi*. W.A. (excluding south-west), N.T., inland S.A.ʼ Vic., N.S.W., Qld; trees, occasionally caves.
Recognition: small brown bat with flattened head and broad nose; forearm less than 1½ inches.

HUGHENDEN BROAD-NOSED BAT, *Nycticeius influatus*. Central Qld (vicinity of Prairie); no habitat recorded.
Recognition: Larger than *greyi*, forearm over 1½ inches, otherwise no prominent features

DOME-HEADED BAT, GOLDEN-TIPPED BAT, *Phoniscus papuensis*. Eastern Qld (Coomooboolaroo Station near Rockhampton, Cape York); no habitat recorded in Australia.
Recognition: muzzle long; ears long, pointed and slightly funnel-shaped with very long and pointed tragus about ½ length of ear; forearm about 1½ inches long.

Group 50 Flying foxes Plate 57

The four of the largest species of Megachiroptera are collectively called flying foxes. Some individuals of the two largest species, the Black Fox, and the Spectacled Fox, have wing spans in excess of four feet, and, when they fly in enormous numbers, as they do from their summer camps, their huge aerial columns are a spectacular sight.

Flying foxes are considered a pest by fruit-growers along the east coast, and, while they are not a serious menace to the industry as a whole, large losses occur on occasions and private gardens, in particular, suffer from the destruction of ripe, and ripening fruit. Flying foxes are principally blossom-eaters and depend on flowering native trees to feed the populations for most of the year.

Where they are causing damage numbers of methods of controlling flying foxes have been tried; those in most use today are gun-club shoots at summer camps and the poisoning of fruit in orchards. The latter method gains its success from the fact that flying foxes seem to lead each other to newly discovered ripe fruit. Through poisoning individuals as they discover ripening fruit, growers can avoid great concentrations of flying foxes which come as the 'news' of a crop spreads.

In the summer, flying foxes congregate in great camps when blossom is abundant and, because of this abundance, they are able to maintain large populations in very small areas. The young are born in the early stages of occupancy of these camps and are looked after by the females for some three to four months. Mating takes place in these camps and, soon after impregnation, a non-gregarious phase seems to start with a segregation of the sexes; this phase culminates in a dispersed and fragmented population which persists throughout the winter months.

Flying fox camps are often very large. Seclusion and shade appear to be the two main factors in their choice as a site, and usual selections are rain forest or mangroves. But when these are not available, ti-tree swamps, the vegetation along creek banks, or even open forest, may be used at times. Tradition also appears to play a strong part in the choice of camp sites and even after long utilization has killed trees, and denuded camps of vegetation, they may still be returned to and used.

Francis Ratcliffe, who was responsible for much of our knowledge of the flying foxes through his pioneering work in the late 'twenties, has estimated that camps contain nearly one-quarter million individuals at densities of between ten to twenty thousand individuals per acre. Generally speaking, the size of camps decreases towards the southern edges of the range of the species. Camps often contain more than one species; for example, the Black Fox and the Grey-headed Fox, or the Red Fox and the Grey-headed Fox are often found in mixed camps in Queensland. No regular seasonal migration of summer camps occurs but local movements between camps are common.

A visit to a flying fox camp is a remarkable experience. There is incessant movement —particularly in hot weather—when they hang, cloaked with one wing and with the other gently waving to produce a current of air. If the camp is awake or disturbed, the foxes start moving and squabbling, and they have no hesitation in taking off and flying to another part of the camp, or even right away.

Francis Ratcliffe has pointed out that these gregarious animals are particularly vulnerable to the reduction and restriction of natural food supplies (i.e. blossoms, wild figs) and the work of Dr John Nelson of Monash University leaves little doubt that the quantity of food within feeding distance of a camp must control the numbers of individuals that can be maintained in that camp through the summer. Accordingly, closer settlement, and clearing of blossoming trees, is bound to affect the density of these very wonderful, gregarious animals.

The Species   Group 50

Flying foxes are unmistakable because of their great size, their dog-like heads, and claws on the second digit of the hand as well as the thumb—unlike microchiropterans which have only clawed thumbs.

RED FLYING FOX, RED FOX, *Pteropus scapulatus* (Plate 57). North-eastern Vic. (has also been recorded at Burnley, a Melbourne suburb), eastern and central N.S.W., Qld (except the south-west), N.T., W.A. (apparently coastal south of the De Grey R.; most southern record Donnybrook, south of Perth); rain forest, sclerophyll forest, sclerophyll woodland, mangroves.
Recognition: reddish-brown or light chocolate coloured; shoulders with orange-brown mantle, often indistinct; long pale yellowish hairs along upper arm under wing membrane.

GREY-HEADED FLYING FOX, GREY-HEADED FOX, *Pteropus poliocephalus*. South-eastern Qld (south of Rockhampton), eastern N.S.W., southern Vic., rare vagrant to Tas.; habitat as above.
Recognition: dark grey, grizzled with lighter hair; shoulders and back of head with reddish-yellow mantle; head greyer than body.

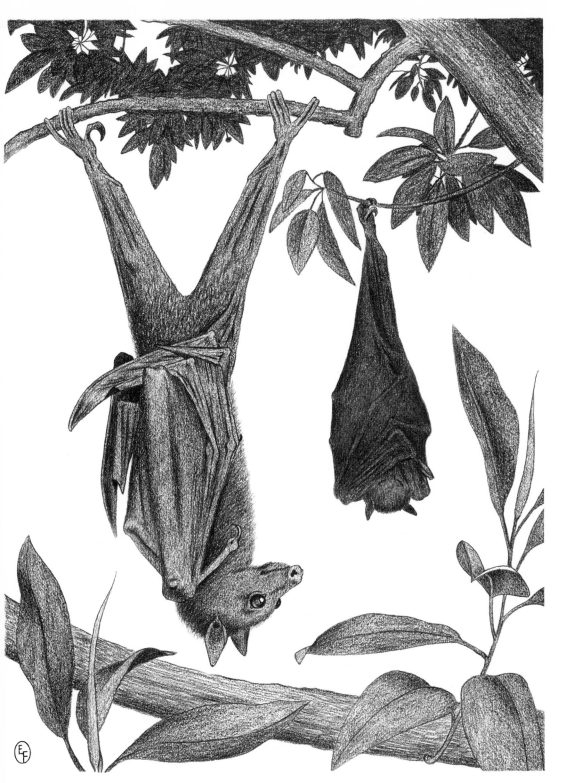

57   RED FLYING FOX, *Pteropus scapulatus*

Group 50

BLACK FLYING FOX, BLACK FOX, *Pteropus alecto*. North-eastern N.S.W., eastern and northern Qld, north of N.T., Kimberley, coastal W.A. (south to Fortescue); habitat as above, preference said to be for mangroves.
Recognition: black or blackish-brown, sometimes somewhat grizzled; shoulders and back of head with reddish to yellowish mantle, mantle occasionally absent.

SPECTACLED FLYING FOX, SPECTACLED FOX, *Pteropus conspicillatus*. North-eastern Qld (from Cape York to Cardwell); preferred habitat rain forest.
Recognition: black, grizzled; shoulders and back of head with yellowish mantle; rings of pale yellow fur around eyes.*

Group 51 Other blossom-eating bats   Plate 58

Five species of blossom-eating bats, as well as those called the flying foxes, have been described from Australia. The largest of these, the Bare-backed Fruit Bat, is easily recognized by its naked back and its habit of hanging by day in caves and mine tunnels. Most megachiropterans hang in trees. The Bare-backed Fruit Bat has only been recorded from Cape York.

The tube-nosed fruit bats, of which two species have been recorded in Australia, are much smaller than the flying foxes, having a wing span in the vicinity of a foot. Little is known about the habits of the tube-nosed bats which are so called because they have remarkable tube-like protruding nostrils. The Queensland Tube-nosed Bat is, like the Bare-backed Fruit Bat, only known from Queensland.

The two remaining species are both very small and could easily be confused with those insect-feeding bats which, like sheath-tailed bats, have dog-like heads; but, unlike them, both of these blossom-eating bats have claws on the index finger as well as the thumb. Little is known of their habits in Australia and it seems likely that they feed on nectar and pollen. The Queensland Blossom Bat has a brush-tongue like a honey-eating bird or the Honey Possum; Dr Nelson who has studied this species in north-eastern New South Wales, suggests that it may form camps like the flying foxes do for a very short period of from two to four weeks in early October.

The Species   Group 51

All species have dog-like heads and, all but the Bare-backed Fruit Bat, have claws on the index finger as well as the thumb.

BARE-BACKED FRUIT BAT, SPINAL-WINGED BAT, *Dobsonia moluccense*. Cape York Peninsula; caves and mine tunnels.

*As well as the four species listed here, another Fruit Bat was collected, at Percy Island in the Barrier Reef, by the naturalist of H.M.S. *Herald* about 1854 during a survey voyage along the Queensland coast. It was thought to be a new Australian species of *Pteropus* and received the name *Pteropus brunneus*. Research by Mr K. Andersen of the British Museum seems to show that it is a specimen of *Pteropus hypomelanus* which occurs in Celebes, around New Guinea, and in the Louisiade Archipelago to the south-east of New Guinea; the specimen was probably a stray into Australia. *Pteropus hypomelanus* has not been identified in Australia since that occasion in 1854 and at present there is no reason to believe that it should be considered a normal part of the Australian fauna.

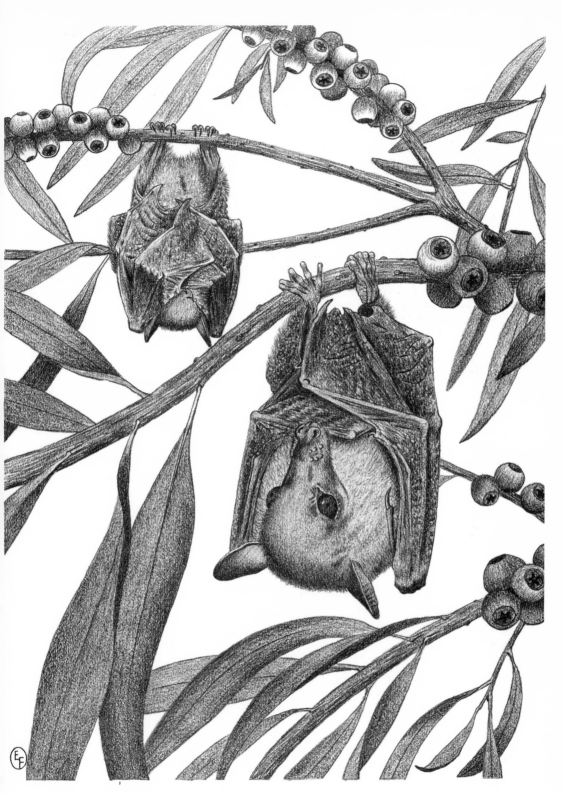

58   QUEENSLAND BLOSSOM BAT, *Syconycteris australis*   Group 51

Recognition: large, forearm about 6 inches long; lower back bare; short free tail; no claw on index finger.

QUEENSLAND TUBE-NOSED BAT, *Nyctimene robinsoni.* Eastern Qld (south to about Brisbane). *Recognition:* projecting tubular nostrils; brown, trace of stripe on back; forearm more than 2½ inches long.

PAPUAN TUBE-NOSED BAT, *Nyctimene albiventer.* Cape York Peninsula, eastern N.S.W. *Recognition:* projecting tubular nostrils; stripe down middle of back not on to head; medium sized, forearm about 2½ inches.

QUEENSLAND BLOSSOM BAT, *Syconycteris australis* (Plate 58). Eastern Qld, N.S.W. (to 28° 20′ S.); vicinity of rain forest and *Melaleuca* swamp. *Recognition:* small, forearm less than 2 inches; pale fawn.

NORTHERN BLOSSOM BAT, *Macroglossus lagochilus.* Northern W.A. (Sunday Island), northern N.T. (Port Essington and Adelaide River); no Australian habitat has been recorded for this species but the specimens taken by Calaby at Port Essington, N.T., were among *Melaleuca* trees on the edge of a fresh-water lagoon close by savannah woodland and a patch of monsoon forest. *Recognition:* like *Syconycteris* but separable from it because the central lower pair of incisor teeth of *Macroglossus* have a wide space between them, but are in contact or nearly so in *Syconycteris.*

11

Non-marsupial carnivores:

THE DINGO AND SEALS

LIKE THE BATS, which flew into Australia, the seals did not find that the water gaps which isolated this continent from the rest of the world were a barrier to their movements. As a result, Australia shares with all the continents around the Southern Ocean the kinds of seals appropriate to her latitudes. The ocean-going earless-seals visit her shores while the much more shore-dependent fur seals and sea lions have settled down and have even evolved two species peculiar to Australia and another which is shared with New Zealand.

The only other native non-marsupial carnivore in Australia is the Dingo, a relative latecomer to the continent. Its presence only serves to highlight the remarkable way in which the composition of the Australian mammal fauna has been controlled by its isolation; elsewhere in the world the Carnivora and their relatives, the hoofed animals, are the most diverse and among the most successful of all mammal groups, and all other continents possess large numbers of species. Yet we have only one native terrestrial non-marsupial carnivore and no hoofed animals at all—and if we are honest, we should admit that we only regard the Dingo as a native because he was here before we, the Europeans, arrived. He, as a species, evolved elsewhere, as we did.

Group 52 The Dingo   Plate 59

The Dingo is still a common animal in most parts of Australia. In its most usual variety it is a tawny-yellow dog with a paler belly and often a white tail-tip and white feet. From the observations of doggers and pastoralists, Dingos mate in April or May, they whelp in the winter between July and September, and the pups are cared for and are trained by the adults until late summer. Each pair of Dingos will have a litter of five to seven pups a year, or even as many as nine, and parents hide the litters in remote and inaccessible places. Normally, the male lives in a separate den from the female and the pups, but both play a part in feeding the pups during weaning and afterwards.

Economically, the Dingo is an extremely important animal because it kills sheep. Its control involves the State Governments in considerable manpower, while the cost to pastoralists of special fencing is enormous. However, before all this cost is blamed on the Dingo, it must be borne in mind, that even where true Dingos are practically absent, domestic dogs running wild may still require similar controls to be exercised. Mr A. R. Tomlinson, the Chief Vermin Control Officer of the Agricultural Protection

Board in Western Australia, has estimated that the total cost of control measures, taken in that State between 1927 and 1954, amounts to several millions of pounds.

One of the biggest needs of the controlling authorities at present is precise knowledge of movements made by Dingos at various times of the year. It is suspected that considerable movements take place and that effective control of Dingos in sheep-growing, and agricultural, areas will not be achieved until reservoirs of Dingo populations outside these areas are isolated or brought under effective control in such a way that Dingos cannot move out from them. Where sheep country borders on cattle country, the problem is particularly acute and it is aggravated by the bonus system. Thus, along the Fitzroy River in South-west Kimberley, the owners of sheep stations are hindered in their attempts at control by the adjacent cattle stations where the attitude is one that dogs are not a menace, and that bonuses are a profitable sideline for station hands and others. Mr Tomlinson records that few of the cattle stations carry out any organized control measures by poisoning, and those dogs which are destroyed are shot or trapped for the bonus. He says: 'As a result, the area might be described as a gigantic natural dog farm.'

In Western Australia between 1927 and 1954 nearly 400,000 scalps were presented for bonus payments. And of those presented in 1954, over half came from the cattle country of the Kimberleys.

It is known that Dingos can be mated with domestic dogs but the extent of cross-breeding which occurs in the wild between domestic dogs and Dingos is a matter of much conjecture among pastoralists and doggers. Dingos are very much like domestic dogs, and it is known that some domestic breeds have Dingo blood by deliberate crossing, but close inquiries and investigations by Mr Tomlinson in Western Australia have failed to reveal any proven instances of crossing in the wild. Professor N. W. G. Macintosh, of the University of Sydney, who has kept Dingos in captivity for more than five generations, points out that a usual argument which is cited as evidence of cross-breeding and racial impurity among Dingos is that, in many populations of yellow Dingos, 'coloured' dogs (including whites and brindles) are often seen. He has found that common yellow Dingos (which presumably had some non-yellow ancestors), bred in captivity for several generations, will produce coloured dogs among their offspring, yet their osteological characters remain uniform. This indicates that it is only in colour that they are diverse, while the constancy of their anatomical type argues against the possibility that non-yellow colour is among new variations introduced among wild Dingos by crossing with domestic Dogs. He points out that a great range of coat colours is recorded by the first European explorers into widely scattered parts of the continent; this indicates that variation in coat colour was widespread prior to the coming of the white man, as it is today.

Professor Macintosh summarizes his work on the Dingo to date by saying that a great number of features separate him from the ordinary dog; for instance, some aspects of his mating behaviour are quite different, his gait is characteristic of neither Domestic Dog, nor Wolf, nor Coyote, and is particularly his own. His howl and the various other noises he makes are not to be matched precisely by any other kind of dog. His reserved and aloof attitude, combined at the same time with a high capacity and apparent desire for affection, is almost unique among the dogs. Anatomically, all features of Dingos are within the range of the various sorts of dogs; nevertheless the average Dingo as exemplified by these characteristics is not matched by any other dog,

59   DINGO, *Canis familiaris*   Group 52

and such tests as have been carried out on the blood serum of Dingos give different results from tests on ordinary dogs.

Very recently Professor Macintosh has examined the skeleton of a Dingo from an archaeological excavation at Fromm's Landing on the Murray River, South Australia. This skeleton is some three thousand years old and it revealed that the morphological pattern of the Dingo has remained constant since that time. Professor Macintosh holds that it refutes the belief that the Dingo population has been changed from its original type by crossing with station dogs.

The Species   Group 52

DINGO, *Canis familiaris* (Plate 59). All Australia except Tas. (nowadays apparently absent from central N.S.W. between the Darling River and Dividing Range, central Vic., and much of the wheat belt and northern agricultural district of W.A.). *Recognition:* See Plate.

Group 53 Marine mammals—the seals   Plate 60

Among the marine mammals which occur in the waters around our coasts only the eared-seals and Elephant Seal give birth to their young on the dry land of the Australian continent and its offshore islands. In the waters about the shores there are also large numbers of species of whales which range in size from the largest animal which has ever lived, the Blue Whale, to the much smaller dolphins. The peculiar-looking Dugong is also a dweller in our coastal waters. Other species of marine mammals, and in particular, the earless-seals of the Antarctic, are only visitors. The whales, and the Dugong, are wholly marine mammals; they never come out on shore and are not included in this book.

Eared-seals appear to be very dependent upon their terrestrial habitat. They can easily be distinguished from true seals (including the Elephant Seal) because they can turn their hind flippers forward and walk on them; they also have small ears at the sides of their rather dog-like heads. Earless-seals cannot turn their hind flippers forward and lie and move with them trailing along behind.

There are three species of eared-seal in Australian waters, the Australian Sea Lion (or White-naped Hair Seal) the Australian Fur Seal and the New Zealand Fur Seal. The last two have been confused in the literature of Australian seals and their present status and distribution has been clarified through recent work in Australia by Miss Judith E. King of the British Museum of Natural History.

The Australian Sea Lion is a hair-seal—so called because soon after it is born it loses the fine dense under-fur which lies under the coarse hair. This under-fur is retained by adult fur seals and makes them valuable to the fur trade. In former times Sea Lions have been killed for their hides and for oil. There is a great difference in size between the bulls and the cows (or clapmatches) and well-grown bulls are the easiest of all Australian seals to identify because of the white or yellowish mane of coarse hair on the nape of the neck; this mane is rather variable and may extend from the crown of the head down the thickly padded neck on to the shoulders. Unfortun-

AUSTRALIAN SEA LION, *Neophoca cinerea*   Group 53

ately, young animals and clapmatches do not possess the yellow-white nape and can easily be misidentified as fur seals, although to anyone familiar with seals, the sharply-pointed noses of fur seals distinguish them from the more dog-faced Sea Lions.

Like all eared-seals, Sea Lions give birth to their young well above the high-water mark and the young are unable to swim until they learn later. Each clapmatch produces a single pup during the breeding season which is between October and December. During this period the seals congregate up above the beaches where each bull defends his small harem of clapmatches until after birth and mating have taken place. Sea Lions are well able to climb sand dunes and even cliffs and Professor Wood Jones recorded an animal as far as six miles inland from the sea. Fish and penguins seem to be the main food of Sea Lions but few studies on the diet of Australian seals have yet been made. Sea Lions are extremely inquisitive animals and seem to be very intelligent. They will investigate swimming humans and will even play with divers.

The true fur seals are smaller animals than the Australian Sea Lion. The males are called wigs and the females clapmatches. These names were used by sealers in the early days. Fur seals seem to prefer rocky inaccessible places on unfrequented islands for breeding places. Like the sea lions they breed in the early summer. In 1964, when the Fisheries and Wildlife Department of Victoria started their current investigations into the biology of the Australian Fur Seal at Seal Rocks in Western Port Bay, the wigs first appeared on the rookeries in November and fought savagely over territories. Towards the end of the month the females began to arrive and pups started to be born. They had all been born by mid-January and the wigs then disappeared leaving the island to the nursing females and their pups which, by then, were beginning to enter the water.

The fur seals, and in particular those of Bass Strait, were formerly the basis of an important industry. Sealing started in 1798 and seal populations were so large that sealing vessels came from all over the world to them. Colonial sealers also operated from Sydney and Hobart and Launceston. As in most early sealing, the methods used were barbarous in the extreme. This particular industry was characterized not only by man's inhumanity to animals, but also to man. It was the local custom of sealers to force Aboriginal women to work and look after them on the islands where they were landed. James Kelley, an early Bass Strait sealer, said that it was their custom for each man to have from two to five native women for their use and benefit.

The industry was short-lived. By 1806 there were signs that the stocks of fur seals were declining. The concept of conservation or controlled cropping was unknown and, by 1860, sealing had almost died out. The most recent sealing in Western Port Bay in Victoria was in 1879; Wood Jones records that the last at Kangaroo Island, S.A., took place about 1880, while V. N. Serventy says that the last sealing party visited the Recherche Archipelago in 1920. Seals are now protected in all Australian waters, but permission is given from time to time for fishermen to destroy individual animals which are interfering with fishing operations by damaging nets, robbing fishermen of catches and by dispersing shoals. The current Victorian investigations of fur seals are aimed to give factual information on the extent of damage done by seals and to give an indication of possible control measures should these prove necessary.

Today, the largest Australian colonies seem to be on Seal Rocks in Western Port Bay where numbers may reach 5,000 and on Lady Julia Percy Island and associated islands off Port Fairy, Victoria, where similar numbers are thought to occur.

Crayfish and fish seem to be the principal recorded items of diet, but on Macquarie Island species of yellow-crested penguins are an important food.

The earless-seals are represented in the Australian fauna by only the Elephant Seal as a resident breeder. The only known breeding colony in historic times was on King Island in Bass Strait but this was exterminated within a few years of its discovery early in the nineteenth century. In recent years, however, there have been records of animals occurring in Australian waters and in October 1958 a cow pupped at Strahan, in western Tasmania. Archaeological records show that some 9,000 years ago the species was common in western Tasmania.

The biology of the Elephant Seal is the best known of that of all Australian seals. This is the result of work which is being done on Macquarie Island and Heard Island as part of the biological research programme of the Australian National Antarctic Research Expeditions (A.N.A.R.E.).

Much is also known of the Elephant Seal in South Georgia where the populations have been exploited under strict control for many years. The present world population of Elephant Seals has been estimated at about 600,000 in mid-year (this figure excludes the fifty per cent of pups that do not survive their first winter). South Georgia contains half of the total world population, Macquarie has about one-sixth, and Heard one-eighth. Elephant Seals come ashore only to breed, moult, and rest. Bulls begin breeding at seven and eight years of age but, at this age they have only limited success in acquiring cows; the older breeding bulls (which are about fourteen feet in length) and which control harems are called beachmasters. All cows are breeding by seven years of age.

Harems of Elephant Seals are very large. On Macquarie, the larger ones have many bulls in attendance; the largest has over 1,000 cows and there are other large harems of between three and six hundred. There are continuous changes in the ownership of harems by beachmasters and, as the harems grow, assistant beachmasters are admitted by the dominant bulls. Harems consisting of less than fifty cows never contain more than one breeding bull at a time, but as they become larger they become impossible for a single beachmaster to control. As other bulls are admitted to a harem, each has his own sphere of influence among the cows. The number of cows to each bull seems to depend upon a number of factors among which is the suitability of the site for defence. Harems were recorded on Macquarie, during 1957, which reached 139, 150, 151 and 184 cows before assistants were admitted by the beachmasters.

In South Georgia, where the population is exploited by sealers, the females become sexually mature at two years of age and bear their first pups when they are three years old. This is in contrast with the situation in unexploited populations. In the latter they first give birth a year later. It is estimated that females in these unexploited populations produce one pup each year and live about twelve years. Bulls may reach an age of twenty years. In unexploited populations bulls do not reach beachmaster status until they are twelve or more years of age but in the exploited populations this occurs much sooner. At birth, pups are about four feet long and are suckled for an average time of twenty-three days; towards the end of this period they are gaining 20 lbs a day while the mother may lose 700 lbs in weight during the suckling period. Mating takes place about eighteen days after the birth of the pup.

The only animals which are apt to be confused with the seals which breed in Australian waters are the species which visit us from the Antarctic, and the Dugong

The latter, although superficially seal-like when it is in the water, belongs to a group of animals quite distinct from seals; this group, which is called Sirenia, contains only one other living species, the Manatee of West Indian waters.

Of the Antarctic seals, only the Ross Seal has never been recorded on our coasts. The Sea Leopard is the most common; one was recorded as far north as Bronte Beach, Sydney, in 1959. The Crabeater Seal has also been recorded near Sydney (at Manly in 1922). But most records of these seals and the Weddell Seal are of sightings along the southern coasts. Dugongs occur in northern Australian waters but they are only likely to be misidentified as seals at the extreme southern ends of their range because there is little geographical overlap. On the east coast, Dugongs have been recorded as far south as Sydney and, on the west, in Shark Bay, but the main populations seem to lie in shallow northern waters between the Townsville area in the east and Shark Bay in the west. Dugongs never leave the sea; they are vegetarian and may be distinguished easily from seals because they do not have separate hind flippers.

Seals are large mammals and in the Australian species the males are always larger than females although this statement does not hold for the Antarctic visitor. Approximate sizes of adults are: Australian Sea Lion, male 10-12 ft, female 8-10 ft; Australian Fur seals, males about 8 ft, females about 5 ft; New Zealand Fur Seals, males about 6-7 ft, females about 5 ft; Elephant Seal, males 18-20 ft (weight about 8,000 lbs), females 10-12 ft (weight about 2,000 lbs). Sizes of the visiting seals are Leopard Seal, 10-12 ft; Crabeater Seal 8-9 ft; Weddell Seal 9-10 ft.

The Species   Group 53

AUSTRALIAN SEA LION, WHITE-NAPED HAIR SEAL, *Neophoca cinerea* (Plate 60). South-western Australia from the Abrolhos southwards (Fisherman Island at Jurien Bay, Carnac, Recherche Archipelago) and along the south coast (islands in the Australian Bight and Spencer Gulf) to Kangaroo Island, S.A. Only known mainland colony at Point Labatt, south of Streaky Bay, S.A.
Recognition: small ear, can turn hind flippers forward, dog-like face; no under-fur when adult; males very large with thick necks and yellowish-white napes or manes; nails of hind feet reaching to level of edge of web between digits; forefoot with first digit longer than second.

AUSTRALIAN FUR SEAL, *Arctocephalus doriferus*. South-eastern Australia from near Port Stephens, north of Newcastle, N.S.W., southwards (Montagu Island, Phillip Island and rocks in Western Port, islands in Bass Strait) to islands around Tasmanian coast. One colony recorded from Tasmanian mainland.
Recognition: small ear, can turn hind flippers forward, short pointed muzzle; nails of hind feet reach well short of level of webbing between digits, first digit of forefoot same length or slightly shorter than second.

NEW ZEALAND FUR SEAL, *Arctocephalus forsteri*. Southern coasts of W.A. and S.A. from Eclipse Island and the Recherche Archipelago in the west, to Kangaroo, Casuarina, Four Hummocks and Thistle Islands in the east.
Recognition: as for the Australian Fur Seal but smaller.

ELEPHANT SEAL, *Mirounga leonina*. Formerly a breeding colony on King Island, Bass Strait, now a regular visitor to Tasmania and, occasionally, to south-eastern Australia.
Recognition: Hind flippers pointing backwards, no ears, very large—any uniformly coloured earless-seal over 10 ft in length is probably an Elephant Seal.

12

The Monotremes

THE EGG-LAYING Monotremes, the Platypus and the Echidna are the two most peculiar of Australian mammals. They are without living relatives outside Australia and there is little knowledge of their fossil ancestry. A group of primitive little mammals called the docodonts which lived in Britain and North America in the Jurassic age about 140 million years ago could be ancestral to them, as could some rather strange fossils represented only by teeth, about ten to twenty million years old, which have been found near Alice Springs. Apart from the presence of giant echidnas (and other monotremes like those alive today) in the deposits of the relatively modern Pleistocene period, of the last million years or so, little is known of the immediate fossil ancestry of monotremes in Australia.

The reproductive organs of monotremes are more primitive than those of any known mammal; unlike others, but like reptiles and birds, monotremes have only a single external opening which combines both sexual and eliminatory functions. They lay eggs, but, like other mammals, they secrete milk to feed their young and have fur and only a single pair of bones in the lower jaw (unlike reptiles which have a series of bones). Their control of body temperature is, like their egg-laying habit, somewhat reptilian.

The platypus is the most popularly known monotreme; it has such an improbable appearance that when specimens first reached Europe, they were thought to be scientific hoaxes made up of parts of birds and parts of mammals.

The most widespread Australian monotreme is the Echidna or Spiny-anteater which is called 'The Porcupine' by many country people and Aborigines.

One species of monotreme which formerly occurred in Australia is now confined to New Guinea and some of the islands near by. This is *Zaglossus*, the Long-beaked Echidna, which is fairly common in Australian fossil deposits.

Group 54 The Echidna   Plate 61

The Echidna is one of the most successful and widely distributed mammals in Australia today, yet it is seldom seen. It has a long beak-like muzzle with a small mouth at the end which opens just wide enough to allow free movement of the long extensile tongue and unrestricted passage to the insect prey on which it feeds. The short limbs terminate in long, strong claws for digging in the soil. One of the claws of the hind foot is particularly elongate; the Echidna uses this for preening between the spines which cover its back and sides. Its under-surface is covered with coarse hair and is without spines. In general body-colour Echidnas range from light brown to almost black, depending on the locality and individual colour pattern. The proportion almost black, depending on the locality and individual colour pattern. The proportion of hair and spines is also variable in the same manner; in Tasmanian Echidnas the hair is long enough to conceal most of the spines.

An adult Echidna is normally over a foot in length and adult males weigh about fourteen pounds. Females are lighter and weigh about ten pounds. Echidnas eat both termites and ants. They also take in a considerable amount of earth along with their diet with the result that the presence of an Echidna can often be recognized by its droppings; these are cylinders of earth filled with the indigestible remains of insects. Moreover, the droppings are not rounded, but, characteristically, are broken off abruptly at the ends.

In a recent study of Echidnas, Dr M. Griffiths, of the C.S.I.R.O. Division of Wildlife Research, has shown that in the vicinity of Canberra, they exhibit very specialized feeding habits. There they open the mounds of the Meat Ant in August, September and October at a time when the virgin queens are present. These queens contain a very high proportion (47.2%) of fat. When an Echidna digs its way out of trouble with its enemies it usually does so by an astonishing vertical subsidence—but in digging out Meat Ants it directs its head and hunched shoulders into the heart of the mound, and, in particular, into the northern side. This is the side on which the ants congregate in later winter seeking warmth. Dr Griffiths suggests that the fat-rich (and energy-rich) diet which they obtain from these ants is important to Echidnas at the time when they are emerging from their periods of winter hibernation.

The Echidna has a pouch which is a deep depression with strong muscular walls. Just how the egg of the Echidna is placed in the pouch, and how long it is incubated there, is unknown, although there is strong circumstantial evidence that the egg is laid direct into the pouch from the vent by curving the body during egg-laying. After hatching, the bristle-less pouch young lives on a very rich thick milk which is secreted on to the surface of the skin on either side of the midline of the pouch. Soon after pointed spines are developed on the back and sides, the young are cast out of the pouch and left to sleep in a den. The mother returns at irregular intervals of one and a half to two days to feed the young one.

Dr Griffiths says that the actual process of suckling is quite touching to watch. The mother, which he had in captivity, would carefully move up to the young one, raise her beak and gently nudge him with it until he lay between her fore limbs, one of the fore paws would then be used to push him well under the body. She would then arch her back keeping the vent well clear of the floor, and the young one would hang on upside down clinging to her abdominal hair with his fore paws. Vigorous audible suckling lasted for about half an hour; there appears to be a well-developed milk-ejection mechanism.

The Echidna is immensely strong, and is an able burrower. When surprised, its immediate reaction is to burrow and because of its shape, its strength, and its spines, an Echidna between rocks is almost impossible to dislodge.

The Species   Group 54

ECHIDNA, SPINY ANT-EATER, PORCUPINE, *Tachyglossus aculeatus* (Plate 61). Throughout Australia and Tas.
Recognition: prickles; long naked tube-like snout.

61   ECHIDNA, *Tachyglossus aculeatus*   Group 54

Group 55 The Platypus

Plate 62

The Platypus is a common animal of the fresh-water streams and lakes in eastern Australia. It is truly amphibious and emerges from its burrow for feeding usually in the late afternoon and early morning—but it is sometimes seen in the water at other times of the day during cold or overcast weather. Locomotive power for swimming is provided by the front feet; the hind feet are used mainly for stability, and trail behind. When swimming submerged, the ears and eyes are closed and the animal swims with its sensitive muzzle close to the bottom. It feeds by putting the muzzle into the mud and gravel. It appears that only the sense of touch is involved in catching prey and only those items actually touched by the muzzle are snapped up. The diet consists of aquatic insect larvae, crustacea, worms, tadpoles and other small aquatic animals. The Platypus surfaces often for breathing and for chewing the prey which is stored in cheek-pouches while it is submerged. Platypuses are often seen swimming along the surface with only the upper part of the muzzle and a small part of the head and body showing.

The most conspicuous feature of the Platypus is its duck-like bill. This is covered with thick, but soft and flexible, naked skin which is a blue-grey colour on its upper surface. A flap of this skin projects backwards from the muzzle over the adjacent fur. Adults have no teeth and use horny pads to crush their food, but true calcified teeth are present in the young.

The feet are webbed and, on the front feet, the web projects for about half an inch beyond the claws. When a Platypus walks on land, or is burrowing, the flap is folded under the palm.

An adult Platypus is normally from one foot six inches to two feet in total length. An average sized male would weigh a little over four pounds.

The burrows of the Platypus are usually about fifteen to thirty feet long and are winding; old burrows may be up to ninety feet in length. The entrance is normally from three to six feet above water-level but during floods the entrance may be submerged. There may also be more than one entrance.

Breeding takes place in spring; from July and August in the northern part of the range, to October in the southern part. The female builds a nest of grass, leaves and other plant material in a chamber at the end of her breeding burrow and she blocks the burrow with earth in one or more places during the period when she is laying and incubating the eggs. The clutch consists of one to three eggs which have a leathery shell and stick together.

David Fleay is the only person who has succeeded to date in breeding the Platypus in captivity, and his observations suggest that the period from mating to egg-laying was probably twelve to fourteen days and the incubation period a further ten to twelve days. The young did not leave the burrow and enter the water till it was seventeen weeks old. Like the Echidna, the female has no nipples and the milk is apparently sucked by the young as it exudes from the ducts of the mammary glands.

The Platypus was formerly hunted extensively for its fur, but for many years now it has been completely protected by law.

Both the male Platypus and the Echidna have a hollow spur on the inside of the ankle which is connected with a gland situated higher in the leg. Although the spur

62 PLATYPUS, *Ornithorhynchus anatinus*   Group 55

of the Echidna does not seem to be poisonous to Man, there are several records of people being injured by Platypuses and extreme caution should be shown in handling them.

Mr John Calaby, of the Division of Wildlife Research, C.S.I.R.O., has recently re-examined what we know of the venom apparatus of male Platypuses and the account here is largely drawn from that work. The poison apparatus consists of a movable horny spur on the inner side of each hind limb near the heel; this spur is supplied from a poison gland situated in the thigh. From the recorded cases, it seems that the effects are not dangerous to man but are painful. When striking, the Platypus drives the hind legs towards one another with considerable force so that the spurs are embedded in any flesh caught between them. All the recorded strikes on humans have been on the hand and the wrist. More or less immediately following injection, in most cases the injured hand began to swell and the swelling extended some distance up the arm even to the shoulder. The swelling subsided in one to a few days. There was immediate intense pain, which lasted up to a day, and the wound felt sore to the touch for some days. In several cases, it was recorded that the victim suffered from sleeplessness, presumably due to the great pain. In at least one case, it is said that the injured arm did not completely recover for some weeks.

The function under natural circumstances of the venom apparatus in the Platypus is not understood. Harry Burrell, who had studied the Platypus extensively and wrote the first important book about them, thought it most likely that it is a weapon of offence employed in attacking other males and natural enemies, but he also believed that the spurs were useful to hold the female during mating. David Fleay saw a Platypus endeavouring to use its spurs on a frog which it eventually ate. He has suggested that the apparatus might also be used to immobilize larger items of prey.

The Species   Group 55

PLATYPUS, *Ornithorhynchus anatinus* (Plate 62). Eastern Qld (south of Cooktown), eastern N.S.W. (westward to about 146°E.), Vic., south-eastern S.A., Tas., (introduced Kangaroo Island, S.A.); fresh-water streams and lakes.
Recognition: thick brown fur; broad duck-like bill.

13

The rare ones

In Chapter 1 we discussed, very briefly, the meaning of the word rarity as it applies to a wild mammal; we saw that, providing the word is taken at its face value (i.e. meaning only that we have few records of the mammal) then it is a useful first step in pointing to the need for a closer examination of the problem. In other words, having found out that an animal is *rare*, we are then bound to go on and find out *why*.

In Chapters 5 to 12 we listed the distributions (and what was known of the habitats) of all the Australian native mammals. We are in a position, therefore, to go through this list and see what is rare, stating, honestly, the criteria upon which each statement of rarity is based. If we do this, we can then go on and determine whether, and how, efforts can be made to find out if particular species are really low in numbers. Then, if we discover that numbers are really low, we can go on to ask why, and whether this betokens danger.

In the list of rare species which follows, a species is automatically included if it has been recorded by not more than ten specimens or confirmed sightings in this century. In addition, some species have been listed which are known from more than ten specimens but are categorized as rare because most specimens were obtained on very few occasions. Examples of such species are Woodward's Wallaroo (*Macropus bernardus*), the Rock-haunting Ringtail (*Petropseudes dahli*), the Yallara (*Macrotis leucura*) and the Desert Bandicoot (*Perameles eremiana*).

Most of the records in the list are of specimens which have been killed. While it may seem strange that our rarest animals are usually only known through their deaths, the reason for this is fairly obvious; it is that most species can only be identified with certainty when an actual specimen has been carefully examined in the laboratory. In fact, it is generally true with mammals, and in particular with the secretive and cryptic smaller sorts, that the presence of an uncommon species is usually only recognized following the accidental killing of a specimen by someone who was unaware of its rarity, or even of its presence. Even when a scientific institution such as a wildlife department or a museum sets out to discover the species of mammals which inhabit a particular piece of country there is seldom an alternative in the first instance to collecting specimens; but once a population of a species has been located an assessment of the density of the population can be made, and its limits ascertained, by techniques which do not require killing. The tragedy which is repeatedly retold in the chapter which follows is that, in almost every case, the discoverers and other interested people have not had the resources, nor the opportunity, to follow up the initial discoveries, so that most populations, discovered through the death of some of their members, have not been investigated and contacts with them have been lost.

Marsupials

Toolache Wallaby, *Macropus greyi* (p. 47): Finlayson reported in 1927 that the Toolache had been seen last in the wild in 1924 at Konetta, twenty-six miles south-east of Robe, South Australia. There have been no confirmed observations since.

Woodward's Wallaroo, *Macropus bernardus* (p. 47): This rock-kangaroo was last collected in 1922 when three specimens were obtained by Mrs P. Cahill at Oenpelli for the National Museum of Victoria. Earlier, in 1918, Mr Cahill presented five specimens to Taronga Park Zoo. Previously it had been collected in Arnhem Land by J. T. Tunney in 1903 and K. Dahl in 1895.

Parma Wallaby, *Macropus parma* (p. 48): The Parma has probably been redis-covered at Gosford, N.S.W. Before this it was last collected wild in Australia at Dorrigo, New South Wales, in 1932. It is established at Kawau Island, New Zealand, where it is numerous (see p. 22).

Crescent Nail-tailed Wallaby, *Onychogalea lunata* (p. 54): It has been rarely collected or reported over the last thirty years. Finlayson records an animal being killed in Central Australia between the Tarlton and Jervois Ranges in 1956, and Butler found the remains of one in 1964 near the Warburton Range, Western Australia. The specimen appeared to have been killed·by a Fox some short time previously.

Eastern Hare-wallaby, *Lagorchestes leporides* (p. 58): Marlow records that the last specimen to be taken in New South Wales was collected in 1890, thirty miles north of Booligal. In 1924, Professor Wood Jones knew of no material collected in then-recent times in South Australia. It was last collected in Victoria (or at any rate on the Murray) in 1857, when it was common.

Hare-wallaby (without name), *Lagorchestes asomatus* (p. 58): This wallaby was described by Finlayson in 1943 from a single skull which was collected between Mount Farewell in the Northern Territory and Lake Mackay on the Western Aus-tralian border by Mr Michael Terry. Nothing more is known of it.

Desert Rat-kangaroo, *Caloprymnus campestris* (p. 67): The Desert Rat-kangaroo is an animal which has been lost, found, and then lost again. In 1931 Mr L. Reese, of Appamunna in the far north-east of South Australia, discovered the first specimens to be seen of the species since it was described by Gould in 1843; and the subsequent locating and redescription of the desert population by Finlayson is one of the classic 'rediscovery' stories of Australian mammalogy. The account is given in Finlayson's *The Red Centre*.

Finlayson found the Desert Rat-kangaroo fairly common on two small flats lying east and west of Cooncheri; from the specimens captured, he concluded that it is probably herbivorous. It builds a small flimsy nest of grass and leaves in a little scooped-out depression, frequently under a small bush. From at least June to December pouch young at all stages of growth are carried.

Unfortunately, no one has maintained contact with the population and when, in 1961, Finlayson surveyed our knowledge of the mammals of the Centre, he could only note that there had been no reliable record since 1935.

Northern Rat-kangaroo, *Bettongia tropica* (p. 68): This species was only recognized as distinct in 1966 by Wakefield and is only known from six specimens altogether. The most recently collected of these was taken in 1932 at Mount Spurgeon, north-eastern Queensland, by Dr P. J. Darlington of Harvard University. It was first collected in 1884 by the Norwegian Dr K. Lumholtz of the University Museum, Oslo, at Coomooboolaroo Station in the Dawson Valley, south-eastern Queensland.

Broad-faced Potoroo, *Potorous platyops* (p. 68): This species was discovered by Gilbert who collected at least two specimens in the vicinity of Goomalling and King George Sound in 1842–3. Subsequently it was collected again by Masters who obtained four in the vicinity of King George Sound and at the Pallinup River. The most recently collected specimens appear to be five which were received from dealers by the National Museum of Victoria in 1874 and 1875. Unfortunately the data with these only indicated that they were from 'West Australia'.

The literature has it that a specimen of the Broad-faced Potoroo was received by the Zoological Society of London at Regents Park, London, in 1908, and that it came from Margaret River in the South West. From what we know of its distribution from sub-fossil remains, this is a most unlikely locality for the species. Mr John Calaby and Mr G. B. Stratton, the Librarian of the Zoological Society of London, have searched for confirmation but have been unable to establish the basis for this identification; moreover, the specimen was not preserved. I am of the opinion that the record had best be discarded; to a person not familiar with small wallabies, it would be easy to confuse a young Quokka with a Broad-faced Potoroo from external features.

Rock-haunting Ringtail, *Petropseudes dahli* (p. 76): Knut Dahl in 1895 found the Rock-haunting Ringtail plentiful at Mary River and at Union Town at the head of the South Alligator River in Arnhem Land. Eight years later Tunney also found it plentiful. However, the only record of this species in Arnhem Land since 1903 is of three specimens from the East Alligator River which reached the National Museum of Victoria in 1912, and two specimens which were collected by the American-Australian Expedition at Oenpelli in 1948. Two specimens were also collected by W. H. Butler at Inglis Gap in the King Leopold Range, West Kimberley, in 1965. These are the first record of the species from Western Australia.

Burramys, *Burramys parvus* (p. 86): Known only from one specimen collected in 1966 at Mount Hotham, Victoria, by Dr D. Shortman and Mr D. Jamieson (see p. 14).

Long-tailed Pigmy Possum, *Cercartetus caudatus* (p. 86): Only eight specimens of this species are known from Australia, all from within fifty miles of Cairns. The first of these was collected in 1908 and the most recent in 1948.

Moonie River Wombat, *Lasiorhinus gillespiei* (p. 94): This wombat is only known from a single restricted locality near St George in southern Queensland. No specimens have been collected since the early part of the century.

Desert Bandicoot, *Perameles eremiana* (p. 100): This bandicoot was collected first by F. J. Gillen of Alice Springs and P. M. Byrne of Charlotte Waters, Northern Territory, in 1896. Specimens have subsequently been collected at a number of localities in the southern part of the Northern Territory, and it was said by Finlayson to have been a fairly common species in 1932–5. It does not seem to have been collected or otherwise recorded since.

Pig-footed Bandicoot, *Chaeropus ecaudatus* (p. 102): The only Pig-footed Bandicoot taken in this century is one collected by John McKenzie on the west bank of North Lake Eyre in 1907. The most recent unconfirmed records seem to be those of Wood Jones, who recorded one being killed (but no confirmatory specimen preserved) between Miller's Creek and Coward Springs to the south and west of Lake Eyre in 1920, and of a skin seen by A. S. Le Souef in 1927 at Rawlinna on the Transcontinental Line. It is not known what became of the skin. Pitjanjarra men in the Musgrave Ranges told Finlayson that they knew of it up to 1926. The most recent confirmed records of the last century seem to be those specimens sent by P. M. Byrne in 1896 to Sir Baldwin Spencer after the Horn Expedition.

Yallara, *Macrotis leucura* (p. 104): This little rabbit-eared bandicoot has a history very much like that of the Desert Rat-kangaroo. It was first described from a much-faded specimen sent, without adequate locality data, to the British Museum in 1887 by the taxidermist at the South Australian Museum. P. M. Byrne, of Charlotte Waters, then collected it in 1896, sending five specimens to Sir Baldwin Spencer. Subsequently, A. S. Le Souef obtained a specimen in 1924, at Mungerani in the Lake Eyre Basin; and L. Ruse and H. H. Finlayson located a population and established the characters of the species, and its variation, from a series of twelve specimens collected in 1931 at Cooncheri on the Lower Diamantina, South Australia. Contact with this population was not maintained and there has been no published record since.

Spiny Bandicoot, *Echymipera rufescens* (p. 100): This species, which occurs widely in New Guinea and the Islands, has only been seen once in Australia. A single specimen was collected, in 1932, by Dr P. J. Darlington of Harvard University, in dense rain forest at Rocky River, in the McIlwraith Range near Coen, Cape York.

Dibbler, *Antechinus apicalis* (p. 118): Three specimens have been collected since 1884, all during 1967 at Cheyne Beach townsite near Mount Many Peaks, Western Australia (see p. 20). All have been collected in a very restricted area and it is not known how far the population extends. Prior to 1884 the species was collected at several widely separated localities, for example, New Norcia, Pallinup River.

Narrow-nosed Planigale, *Planigale tenuirostris* (p. 120): Described from a specimen from between Bourke and Wilcannia on the Darling in 1928, a second specimen was sent to the Australian Museum in 1945 by Miss June Kirkby of Bellata in the Moree District, New South Wales. Miss Kirkby said that the species was common at that time. Since then two records have been published (Cunnamulla, Queensland, and Nyngan, New South Wales) and it has also been collected at Abydos Station in the Pilbara of Western Australia.

Kimberley Planigale, *Planigale subtilissima* (p. 120): Until 1949 this species was known only from a single immature specimen which had been collected by the Swedish Mjöberg Expedition in 1913 at Noonkambah on the Fitzroy River in the Kimberley, Western Australia. In 1949 B. Rudeforth of the University of Western Australia collected six specimens at the Kimberley Research Station on the Ord River near Wyndham. Rudeforth attempted to breed them but without success.

Long-tailed Dunnart, *Sminthopsis longicaudata* (p. 122): This species is only known from four specimens. The original was collected by G. A. Keartland in the Pilbara, Western Australia, prior to 1908. There is a second specimen without any locality apart from the vague 'Central Australia' in the National Museum of Victoria. A third was presented to the Western Australian Museum in 1940 by R. N. W. Bligh; it was collected at Marble Bar. The fourth specimen (little more than a broken snout) is in the same museum; it was found by A. M. Douglas among the chewed-up remains in a Ghost Bat roost, also near Marble Bar.

White-tailed Dunnart, *Sminthopsis granulipes* (p. 124): The White-tailed Dunnart was collected first by G. Masters in the collection that he made between 1868 and 1869 at 'King George's Sound and Salt River'. Between 1925 and 1965 nine specimens have been collected (all 'accidentally'*) at various localities in the drier inland periphery of the South West of Western Australia.

Sandhill Dunnart, *Sminthopsis psammophila* (p. 124): Only a single specimen of this, the largest of the *Sminthopsis*, has ever been recorded. It was collected by the Horn Expedition in 1894 near Lake Amadeus in Central Australia.

Hairy-footed Dunnart, *Sminthopsis hirtipes* (p. 124): Eight specimens are known; two of these were collected at Charlotte Waters, Northern Territory, one between Mount Farewell and Lake Mackay on the Western Australian border in 1933, and the remainder between 1930 and 1967 in desert areas of Western Australia (Canning Stock Route, Great Victoria Desert, Elder Creek at the Warburton Mission, Neale Junction north of the Nullarbor).

Thylacine, *Thylacinus cynocephalus* (p. 130): Once common enough to have a bounty placed upon it, the last definite record of a specimen being killed in the wild was in 1930. The last captive animal in Hobart Zoo died in 1933. Tracks and sheep-killing in the Derwent Valley near Hobart in 1957 have been identified with the Thylacine by Dr E. R. Guiler, the Chairman of the Animals and Birds Protection Board, Tasmania. In 1966 the Board informed the International Union for the Conservation of Nature and Natural Resources that 'Reasonably reliable sightings have been made in widely scattered localities such as the Cardigan River on the Queenstown Highway, the far North West Coast, the Tooms Lake region and the Cradle Mountain-Lake-St Clair National Park. Thylacines may also occur in Frenchman's Cap National Park.' Dr Guiler informs me that a lair was discovered on the west coast in 1966.

---

*This term covers capture by cats, children, farmers, and by all other agents who are not collecting deliberately. It is fortunate that many mammals caught in this way are sent to museums for identification. If they were not, our records of many rare species would be even more sparse than they are.

Native rats and mice

False Swamp-rat, *Xeromys myoides* (p. 140): Seven specimens of the False Swamp-rat are known. The first was received at the British Museum from the Dresden Museum in 1887; it was from Mackay, Queensland. Subsequently J. T. Tunney collected another specimen on the Alligator River, Northern Territory, in 1903 (but this was not recognized until identified by J. A. Mahoney in 1965—the skin was sent to London soon after its collection and lay there misidentified among the mice; the skull was in Western Australia, without locality data beyond a collector's number, which showed, beyond doubt, that it belonged to the skin which Mahoney found by searching in London. The skin had both collector's number and locality label). Five specimens were collected by W. A. McDougall, prior to 1941, during research on cane-field rats, at Mackay, Queensland.

Big-eared Hopping-mouse, *Notomys megalotis* (p. 146): The only localized specimen of this species was collected by John Gilbert on 19 July 1843 at the Moore River, in the vicinity of New Norcia, Western Australia. Another specimen purchased from Gould by the British Museum is merely stated to be from Australia.

Darling Downs Hopping-mouse, *Notomys mordax* (p. 146): This species is known from a single damaged skull which was received at the British Museum with part of the Gould Collection. The details of its collection are unknown beyond the fact that it was stated to have come from the Darling Downs among material sent to Gould by his brother-in-law, Charles Coxen. It was registered in the British Museum Collection in 1846.

Long-tailed Hopping-mouse, *Notomys longicaudatus* (p. 146): This was first collected at New Norcia, Western Australia, by John Gilbert and has not been seen since in Western Australia. However, numerous specimens were collected in Central Australia on, or immediately after, the Horn Expedition in 1896, and by Spencer and Gillen in 1901 at Barrow Creek. It has not been collected since. Finlayson in 1961 wrote that he had not been able to collect further material, and reports of the species in the Centre are only vague.

Short-tailed Hopping-mouse, *Notomys amplus* (p. 146): This species is only known from two specimens which were among those resulting from the Horn Expedition in 1896.

Macdonnell Range Rock-rat, *Zyzomys pedunculatus* (p. 148): This fat-tailed rock-rat was first made known from six specimens collected in the vicinity of Alice Springs, and another from Illamurta in the James Range, which were described with the results of the Horn Expedition, in 1896. The only records for this century are by Finlayson, who notes an unstated number of individuals (which he had not examined himself) at Hugh Creek in 1935, Napperby Hills in 1950, and Davenport Range in 1953.

White-footed Tree-rat, *Conilurus albipes* (p. 142): Very little indeed is known of this species. There are two specimens in Australia and several in European museums, all collected in the last century, Gould said that it extended from southern Queensland (Darling Downs) to Victoria.

White-tipped Stick-nest Rat, *Leporillus apicalis* (p. 144): This was common along the Murray and Darling Rivers in 1856–7, but there seem to be no records of it in that area since. The most recent records are of two specimens collected in 1933, by A. Brumby of Ernabella, west of Mount Crombie, south of the Mann and Musgrave Ranges, in South Australia. Finlayson says that in 1941 he had reliable information of its presence west of Mount Peculiar.

Hastings River Mouse, *Pseudomys oralis* (p. 155): As far as is known, this species has only been collected twice; both specimens were taken in the first part of the last century. The first of these was purchased by the British Museum in or prior to 1847, from a dealer named Pamplin. No precise locality was given for it other than 'Australia'. The other specimen which was from Hastings River, New South Wales, is in the Gould Collection in the City of Liverpool Museum, England.

Alice Springs Mouse, *Pseudomys fieldi* (p. 155): There is some doubt as to whether this species really exists because it is possible that its distinguishing features are merely those of an aberrant individual specimen of another species. Comparison is made particularly difficult because its skull is very badly damaged by being crushed. The unique specimen was collected at Alice Springs in June 1895.

New Holland Mouse, *Pseudomys novaehollandiae* (p. 154)*: Various specimens have been allocated to the New Holland Mouse by different authors. It seems, however that the only material which can now be definitely allocated to it was collected in the early part of the last century and seems to have belonged to the Gould Collection. The original specimen came from Yarrundi on the upper Hunter River, New South Wales; other specimens in that collection are from unknown localities in New South Wales.

Smokey Mouse, *Pseudomys fumeus* (p. 154): This species was not known until 1933; and up to 1962 only four specimens of it were known, all from the dense, scrubby, forest country in the Otway Ranges, Victoria. In 1962 Mr R. Warneke and his team of the Fisheries and Wildlife Department, Victoria, located it in the Grampians. Altogether, sixteen specimens are now known, including those from the Otway Ranges and two localities in the Grampians; its status seems to be quite good in the

*Early in 1968 J. A. Mahoney identified as *Pseudomys novaehollandiae* a small mouse which had been collected about 6 December 1967, in Ku-ring-gai Chase National Park, near Sydney, by Ranger Mr G. Spencer and sent to the Australian Museum for identification. In February 1968 Mr Kent Keith of the Division of Wildlife Research, C.S.I.R.O., discovered that the animal was abundant in a limited area on the southern side of Port Stephens, N.S.W.; there it inhabits a regenerating stage of the understorey of dry sclerophyll forest which follows burning. At that particular stage of regrowth, the seed of *Acacia* and other seeding legumes is abundant. For full information *see* Mahoney, J. A. & Marlow, B. J. (1968). The rediscovery of the New Holland Mouse. Australian Journal of Science, Vol. 31, pp. 221-3, and Keith, K. & Calaby, J. H. (1969) The New Holland Mouse, *Pseudomys novaehollandiae* (Waterhouse), in the Port Stephens District, New South Wales. C.S.I.R.O. Wildlife Research, Vol. 13, pp. 45-58.

Grampians, but is not yet known elsewhere and it has not been relocated in the Otway Ranges.

Western Mouse, *Pseudomys occidentalis* (p. 154): The Western Mouse was first described as new in 1951 from a specimen collected by J. Baldwin in 1930 at Tambellup, Western Australia. Two specimens have been collected since, one from Hatter's Hill, near Lake Grace, Western Australia, in 1933, and another from eight miles south of Nyabing, Western Australia, in 1953.

Bats

Little Northern Territory Long-eared Bat, *Nyctophilus walkeri* (p. 165): The only certainly identified specimen of this bat is very small. It was collected in 1891 by Commander J. J. Walker of H.M.S. *Penguin* at the Adelaide River, Northern Territory. Another specimen, thought possibly to belong to this species, has been recorded by J. L. McKean and W. J. Price in 1967.

Warty-nosed Horseshoe Bat, *Hipposideros semoni* (p. 166): Few Australian specimens of this bat are known. The original specimen came from Cooktown prior to 1903. In recent years, the American Museum Archbold Expedition to Cape York in 1948 collected seven specimens north of Cooktown and B. J. Marlow, the Curator of Mammals of the Australian Museum, collected one at Coen in 1960. The species also occurs in New Guinea.

Lesser Warty-nosed Horseshoe Bat, *Hipposideros stenotis* (p. 166): The first two specimens of this species were collected by Knut Dahl of the University Museum, Oslo, on the Mary River, Northern Territory, in May 1895. There the species was not uncommon; but it has only been collected on three other occasions. There is a female and a juvenile from the King River, Northern Territory, in the National Museum of Victoria, a specimen from Mount Isa in the Australian Museum, and one from Derby in the Macleay Museum of the University of Sydney.

Norfolk Island Scurrying Bat, *Tadarida norfolkensis* (p. 168): For many years this species has been confused with eastern Australian specimens of the Little Flat Bat, *Tadarida planiceps*. But now, if we follow Dr Felten's concept of that species, only one specimen, said originally to have come from Norfolk Island, is known of *Tadarida norfolkensis*. It was collected before 1838.

Troughton's New Guinea Sheath-tailed Bat, *Taphozous mixtus* (p. 170): This species occurs in New Guinea, but the only Australian records are those of three specimens collected in 1948 by the American Archbold Expedition to Cape York, at Browns Creek, Pascoe River.

Yellow-headed Pipistrelle, *Pipistrellus javanicus* (p. 176): Nothing is known of the origin of the only two specimens of this species which have been recorded from 'Australia' beyond that they were presented to the British Museum prior to 1878, by the Earl of Derby. The species has a very wide distribution outside Australia.

Timor Pipistrelle, *Pipistrellus tenuis* (p. 176): Like the preceding species, this bat has a wide distribution outside Australia; the only Australian specimen known was collected by Knut Dahl at Roebuck Bay, near Broome, Western Australia in 1895.

Small-footed Myotis, *Myotis australis* (p. 177): This species was described in 1878 from a specimen in the Leyden Museum, Holland; the specimen was said to have come from New South Wales.

Hughenden Broad-nosed Bat, *Nycticeius influatus* (p. 177): The specimen upon which this species is based was collected at Prairie, near Hughenden, Queensland, in 1923, by Captain G. H. Wilkins. It is not yet certain that the species is fully distinct or whether it is merely a very large specimen of *Nycticeius greyi*—a species which is known to have a rather wide size range. However, two additional specimens, one from central Queensland and the other from south-eastern Queensland, have been collected since, which are very nearly as large as the original. Accordingly, I list it here as a distinct species only known from one, or possibly three, specimens.

Dome-headed Bat, *Phoniscus papuensis* (p. 177): Five specimens of this New Guinea species have been collected by the Norwegian, Dr Carl Lumholtz, at Coomooboolaroo Station, near Rockhampton, Queensland, in 1884; three others were collected in 1897, at Cape York, and another at Cloncurry some time before 1929.

Papuan Tube-nosed Bat, *Nyctimene albiventer* (p. 182): This species has only been taken in Australia on three occasions. In 1912 Dr K. Andersen recorded it from Cape York and from various localities in New Guinea and the Islands; prior to 1908 a specimen was collected at Cooktown, Queensland, and another at Wee Jasper on the Goodradigbee River, New South Wales, some time before 1922. Both latter specimens are in the Australian Museum.

Northern Blossom Bat, *Macroglossus lagochilus* (p. 182): This tiny blossom bat has been recorded from Australia in this century on four occasions—at Sunday Island, Western Australia in 1911, where one specimen was collected by the Swedish Mjöberg Expedition; on Cape York Peninsula in 1948, where three specimens were collected by the American Archbold Expedition; at Port Essington in 1965 where four specimens were collected by John Calaby of the Division of Wildlife Research, C.S.I.R.O., and in 1967 when two specimens were collected by the Animal Industry Branch of the Northern Territory at the Adelaide River, Northern Territory. The species is widely distributed outside Australia from Celebes to the Solomon Islands.

14

Those with changed status

A LIST of rare species like that in the last chapter is a useful pointer to one direction for investigation but, if we pause to think about it, there is another category which should be examined before we can say that the names of all the species which are possibly in danger of extinction are presented for closer scrutiny.

Among those species not included with the rare ones are some which, while they are not rare by our criteria, warrant close examination. These are species which were formerly common but are no longer so common, or they are species which are now only common in a fragment of their former ranges.

These fresh categories are difficult to evaluate because they depend upon information as to past abundance. These kinds of information are difficult to obtain and assess because they are usually highly subjective in nature. If such information is not interpreted correctly it can lead to a very wrong idea of the status of a species. For example, the native-cat called the Chuditch (*Dasyurus geoffroii*, see p. 108) is very rare in all parts of its known range except in the south-west of Western Australia where it is very common—but, on examining records, we will see that there is no evidence that the Chuditch was ever anything but rare outside the South West. Compare this with the Rabbit-eared Bandicoot which was formerly very common in the South West, in the wheat belt of Western Australia, in the more fertile parts of South Australia, and was widespread in inland New South Wales. Outside these areas, we have little real information about its previous abundance. At present, we know that it is absent from the places where it was formerly common and it is only patchily distributed as sparse populations over the rest of its wide range where this is outside agricultural districts. The status of such a species clearly requires close examination and must give considerable concern until it is understood.

In presenting a list of species which appear to have changed status, we must bear in mind that the historical data upon which the suspected change depends are not necessarily reliable and the list should be treated with extreme caution. Nevertheless, like the list of rare species, it is a useful starting point for further examination—but in the case of these species there is very great urgency to find out what has occurred and, if possible, why.

Marsupials

Quoll, *Dasyurus viverrinus* (p. 108): Common in Tasmania but apparently restricted to small pockets on the mainland.

Numbat, *Myrmecobius fasciatus* (p. 128): Except for a record in the Everard Ranges by Finlayson, there have been no records outside Western Australia in this century.

Marl, Barred Bandicoot, *Perameles bougainville* (p. 100): Common today on Bernier and Dorre Islands in Shark Bay. Otherwise on the mainland it has only been recorded in recent times from small pockets in Victoria. Previously not rare on the mainland.

Rabbit Bandicoot, *Macrotis lagotis* (p. 104): Formerly very common in agricultural districts of Western Australia, South Australia and New South Wales; now apparently absent from these places, but patchily distributed in non-agricultural parts of its former range.

Queensland Hairy-nosed Wombat, *Lasiorhinus barnardi* (p. 94): Formerly occurring in both central Queensland and the Riverina district, New South Wales, it now appears to have gone entirely from New South Wales and only a small population is known in Queensland. The closely related *L. latifrons* is still common in parts of South Australia, but in restricted localities.

Tasmanian Bettong, *Bettongia gaimardi* (p. 67): Common in Tasmania, it has not been collected on the mainland for many years. Probably the most recent mainland record is from the Grampians in 1910, but there is no confirmed record from the mainland in this century.

Boodie, *Bettongia lesueur* (p. 68): Common on Bernier and Dorre Islands in Shark Bay, Western Australia and on Barrow Island. It has not been recorded on the mainland of Western Australia for many years, yet it was extremely common there up to the 1930s. At one time it occurred in the Northern Territory, New South Wales and South Australia. Finlayson records that it persists today only as a rare form in Central Australia in the drainage of the Sandover and Plenty Rivers.

Woylie, *Bettongia penicillata* (p. 67): The Woylie is common in the Wandoo forests of the South West, but is very uncommon elsewhere in its former wide range across southern Australia.

Quokka, *Setonix brachyurus* (p. 53): Up to the early 1930s the Quokka was abundant in the wetter parts of south-western Australia but it occurs there today only as small scattered colonies. It is abundant on Rottnest Island and Bald Island.

Banded Hare-wallaby, *Lagostrophus fasciatus* (p. 58): The Banded Hare-wallaby is common on Bernier and Dorre Islands in Shark Bay, but no specimen has been taken on the mainland of Western Australia since 1906. It may never have been very common on the mainland but it probably occurred in large numbers in very local situations.

Bridle Nail-tailed Wallaby, *Onychogalea fraenata* (p. 54): This wallaby is said to have been formerly very common in inland New South Wales but there has been no reliable record of it for thirty years. The most recently taken specimens of it are from Manilla, New South Wales, where it was collected in 1924. Two specimens were taken at the Dawson River in south-eastern Queensland in 1929, and at an unrecorded place, possibly in central Queensland, in 1937. Troughton in 1941 recorded that an attempt had been made to preserve the species by introducing it into Bulba Island, Lake Macquarie, New South Wales, but a recent inspection by Mr F. Hersey, the

Chief Field Officer of the New South Wales Fauna Protection Panel, has failed to reveal the presence of any.

Rats and mice, and bats

So little is known of the former abundance of most of the rodents and bats that none of them can be included here with any certainty. However, it is probable that the Stick-nest Rat and Gould's Native-mouse should be included.

Stick-nest Rat, *Leporillus conditor* (p. 144): This rat was once common along the Murray and Darling Rivers but no specimens have been recorded from there since 1857. In this century material was collected in 1907 near the western shore of Lake Eyre North and then, later, in 1922 by Ellis Le G. Troughton at Fisher and at Ooldea on the Transcontinental Line. Apart from a report that it was rare in the Lake Eyre District by 1931, it has not been collected or recorded on the mainland since then. In 1920, however, Professor Wood Jones located a population (which some regard as a separate species) on Franklin Island in the Nuyts Archipelago, South Australia. This insular population appears to be flourishing and secure.

Gould's Native-mouse, *Pseudomys gouldii* (p. 154): This species was collected in the last century from widely separated localities, but in recent years has only been recorded at Rawlinna and Ooldea (both on the Transcontinental Line). One specimen in the National Museum of Victoria seems to have been collected in 1857 by William Blandowski during the expedition which took him some 300 miles up the Darling from its junction with the Murray. Another two are said to have been collected north of the Hunter River, New South Wales, prior to 1855 and are in the Gould Collection; Mahoney is very doubtful of this locality on the grounds that its remains are apparently absent from recent bone deposits from New South Wales east of the Divide. Gould notes that two specimens were sent to him by Mr Strange (his collector in South Australia) and were from between the Coorong and Lake Albert.

15

The how and why of conservation

IN THIS BOOK I have attempted to outline the knowledge that we have of our mammals and to bring to light the problems connected with their conservation. How much more we could do to conserve the mammals if only we knew which species really needed conservative action and how we should go about doing it.

If conservation were only an academic problem, in which time were not a factor, a straightforward plan could be produced which we could develop in good time. First, we would survey to see what lived where; from this we would discover the distributions of all species, their habitats, and their abundance.

Secondly, we would gather biological data on each species, including the size of its populations, the rate at which it breeds, what it feeds on, what are the things which interfere with its mating (and these include all kinds of psychological and physiological factors which determine whether it feels like mating or not), what are its particular requirements for shelter, what needs it has for living room and the peace to go about its own business in the manner and at the time of day it requires.

Having got this we would be in a position to reserve because we would know what physical space and ecological requirements were necessary for each species and, accordingly, could select areas to meet these requirements.

Finally, we would design management programmes for each species and for each reserve to make sure that the populations continued to get what they needed to provide for their permanent conservation.

All this would be fairly straightforward. It would take a great deal of effort and a great deal of time—but time is not on our side in this matter. Even in the south-west of Western Australia alone, a million acres of new land each year has been brought into production since the end of the war, and similar things are happening all over Australia. This cannot continue indefinitely without wholesale extinction.

Those of us who love nature know that the changes that we are bringing to the countryside are so great that all but a few species which have learned to get on with man (as the Brush Possum of our suburban gardens has done) must be seriously affected; this is certainly true of the animals of urban and agricultural areas. Even if most species are not yet threatened with extinction, almost all of them have drastically reduced ranges. In the pastoral lands of the inland north, the effect is less obvious; but it is still there as a result of disturbance by introduced animals and through selective grazing by stock and the overfrequent burning which are changing the botanic composition of the countryside. As a result, the environment which gives both food and shelter to the animals is being changed.

As a result of this urgency, it is obvious that we must design a crash programme that will allow us to take short cuts. Obviously, the first way to do this is to concentrate our effort on those species which we know to be rare (Chapter 13) or appear to have changed in status (Chapter 14). Let us through survey of the rare ones evaluate rarity to see what it really means. Are these mammals really absent, or can we find them before the bulldozer, the scrub-roller, and the stock get there first?

In the case of those where we suspect a change in status, we must collect biological data and see if we can find out why the suspected change has occurred and whether we can do something to prevent it from going farther.

Finally, because we have quite a lot of information on distribution and habitat of almost all species (although it is of a rather sketchy kind) we can make an attempt to secure a reserve for each, even if we cannot be certain that the reserve will be adequate.

'But', you say, 'all this is work for professionals; it is going to involve a lot of field work, and many experiments with captive animals—and it is going to cost a packet. What can an ordinary person like me do? Does it mean that conservation in this modern world is so technical that the man in the street cannot help except through giving money, and through not killing individual animals?'

What can you do?

1. The first thing that you, a nature lover or a responsible Australian, can do at this stage is to help the various species to gain time. Use your influence as a citizen to put off the *total* destruction and alteration of habitat so that, throughout our continent, there are representative areas of all the different kinds of animal and plant communities which are sufficiently large to maintain their own populations of wild things secure from interference. You should bring pressure to bear on legislators to ensure that these reserves and sanctuaries are secure from interference by man and his introduced animals; some of them should be large enough to provide recreational facilities for human holiday-makers at a level which will not interfere with the wild communities. Inside these areas the species will survive and the biologist and naturalist will study them. Support and understand the work of National Parks authorities who manage reserves and the work of the biology departments of your universities, because they are the people who will gather biological data and interpret it.

As a result of their studies, information will come which will allow the reserves to be managed in the best interests of conservation; in addition, knowledge will be obtained which we can use to make it possible for some, at any rate, of the native species to continue to co-exist with us in the places which we must develop for production.

2. The second thing that you can do is to support the Australian Conservation Foundation,* and other naturalist and conservation societies in your own State. Because they represent a unified voice of naturalist opinion, local societies have an important part to play in moulding public opinion about conservation and are often the most successful local watchdogs against the alienation of fauna and flora reserves as the result of local pressures.

3. The third thing that you can do is to join with us in the work of survey. Australia is a huge continent and, no matter how desirable it would be, professional biological surveyors are not sufficient to do the job. Small mammals are being killed by accident

*Write to the Foundation at Box 804, P.O. Canberra City, A.C.T. for information about its aims and programmes.

all over Australia—some by cats and others during clearing or accidentally in traps, post-holes and so on. Do not waste these. Each mammal sent in to a museum is a step in the survey. Interest your friends in mammals, get them to send in information, and if you do not get the response from your museum which you think desirable, agitate until it is equipped and staffed well enough to identify accurately and to survey energetically. Much survey in Australia to date has been left to overseas workers: this is ridiculous. It is the job of your museum.

Ask what the Commonwealth is doing in this field. Why is there no Biological Survey of Australia or Museum of Australian Biology? The Division of Wildlife Research of C.S.I.R.O. is doing what it can, but this is not its real task. Surveys must be carried out and there is little time. The Commonwealth should take part in it.

4. The fourth thing that you can do is to build up the State Authorities which are responsible for conserving the mammals of your State. Ask whether the Authority has the money to be staffed and run by adequate numbers of professionally qualified ecologists and trained conservationists. Do not be satisfied with administrators and wardens, no matter how sincere they are. In your Health Department you insist that there are adequate numbers of physicians and surgeons—men who by their training, are qualified to diagnose and prescribe. Do you apply to your National Parks Authorities and your Fauna Department the same criteria? If, in your State, you do not, agitate until you do. Look at the Laws as well. Are they helping conservation, or are they merely regulations which protect individual creatures but do little to secure the conservation of environments and populations? Such laws can destroy public interest and co-operation, and even hinder survey and the gathering of the biological data upon which conservation will depend. And at the same time they may be inadequate to prevent the destruction by mining, or other activities, of the environment of reserves.

5. Next, interest yourself in those native species which are being exploited commercially. Remembering that we know very little of their populations and the level of predation they can stand, it is of vital importance that cropping should be closely monitored so that close estimates can be worked out of the size of the populations and their permissible cropping rates. Every natural population which comes under a new and unnatural form of mortality ought to be regarded as a population in danger until such rates have been calculated and are being closely watched. Who is doing this in your State? If there is insufficient public money to provide the expert staff to do this, ask why the industries themselves should not be made to pay for it. It is only right, too, that the same humanity and attitude to waste which we apply to our domestic animals should be enforced in these industries. As we know only too well, the cropping of wild populations is often accomplished by cruelty, brutality, and, at best, unnecessary callousness. We must insist that the methods of such industries are very strictly controlled.

6. Finally are we making full use of the land which we already have in production? There is a great temptation to go on clearing land when the productivity of already-cleared land is not at its maximum. The employment of natural bush for recreation and conservation is a valid form of economic land use. At any rate, it is wasteful to misuse it for other purposes when land already alienated is not being used to its full.

Why conserve wildlife?

In this modern utilitarian age why should we bother about conserving at all?

This is a question which taxes every mature and thinking naturalist. The use of land for reservation means taking it out of other forms of production which are more closely related to human needs; if, in this time of human population explosion, we are to do this, it must be done in a responsible and clear-thinking fashion.

The argument for conserving soils, or forests of trees with an economic value, is easy to put. But the conservation of wildlife and native bush without an obvious market or economic value is more difficult to support. What, then, are the arguments? Why conserve wildlife?

First of all, there are a few immensely practical reasons as to why conservation is an important economic asset. In countries with advanced economic development and dense human populations, the need for recreation away from the complexity of modern machine-age life is becoming increasingly important. In addition, the presence of National Parks often brings financial returns to surrounding communities which may even be greater than the return which one could get from them as pasture or crop. The possibility that parks will help local economy may be added to the incalculable benefit gained by those who visit these parks for rejuvenation and refreshment.

The other important practical asset which natural communities of wildlife give us is that they provide the only example of the way in which nature herself, through millions of years of evolution, has produced the balanced natural communities which are perfectly adapted to the particular physical and climatic conditions peculiar to each area. These natural communities are a yardstick against which developers of similar country can measure their own efforts, and by so doing, understand nature's way of solving the problems which they themselves face in utilization of the land.

Conservation is also an important part of education. In learning to understand and appreciate the wild things, our young people are encouraged to search for themselves, to respect values beyond their own immediate needs, and to appreciate the social codes which place value on the protection of the weaker by the stronger. In our human world of real or of simulated violence in daily life (in the make-believe of films and television) the balance provided by the serenity of nature is of very real importance to our growing nation.

A few generations ago, the answer to the question 'Why conserve?' would have been easier to give to a society much more receptive to authoritative statements. At that time the argument adequate to satisfy most people could have been: 'These are God's creatures. What right have we to destroy them?' Today, with our much greater demand for reasons based upon material, scientific and logical processes, we might argue instead 'Each species is the product of over six hundred million years of evolution, and we ourselves are one of these. In destroying any one of these species, we are destroying a facet of our potential understanding of the world about us—we owe a responsibility to each other to see that this does not happen.'

Why conserve? Why paint? Why listen to good music? Why read books? These are the aesthetic sides of human activities, the things which take man out of himself and set him above the other animals.

Appendix I Suggestions for further reading

Without help it is not easy for the beginner to find his way about the modern, highly specialized literature of Australian mammals. This is because most of the precise information of value to him, once he wants to go beyond the stage of general information, is hidden in the enormous volume of papers appearing in a multiplicity of professional journals. The problem is to know where to begin. In an attempt to help those who want to find out more, a number of references are listed in this chapter under the headings used in other parts of the book. Consulting them, and the list of additional references which most of the papers contain, will give a lead into the subject.

In presenting this list I have not abbreviated the names of journals in the rather forbidding style common to scientific works. However, when the bibliographies of the papers listed here are consulted, such abbreviations will be encountered. Do not be put off by their unfamiliarity or by that of the names of apparently obscure journals in the list. Most librarians can obtain them with little difficulty if they are asked.

The articles referred to here vary widely in the level at which they are pitched. Some of them refer to accounts in natural history magazines or books which are aimed at the general reader, yet others are to papers in which professional zoologists have recorded the results of their original research. But, in compiling the list, I have tried to favour papers which use ordinary language and I have avoided papers which merely deal with problems of classification and identification; in this way I have tried to make it an introduction to original observations on the lives of animals and their behaviour. The list does not include the common, easily-available books on the natural history of Australian mammals except where they are by an original research worker and are a valuable source of information on some specialized subject.

Australian mammals today

C.S.I.R.O. (1960): The Australian environment. 3rd revised edition. Melbourne: C.S.I.R.O./ Melbourne University Press. 151 pp.

KEAST, A., CROCKER, R. L. & CHRISTIAN, C. S. (1959): Biogeography and Ecology in Australia. The Hague, Junk: 640 pp.

RIDE, W. D. L. (1968): On the past, present and future of Australian mammals. Presidential Address Section D: Zoology. Australian Journal of Science, vol. 31, pp. 1–11.

A second chance

FINLAYSON, H. H. (1932): *Caloprymnus campestris.* Its recurrence and characters. Transactions of the Royal Society of South Australia, vol. 56, pp. 148–67.

FINLAYSON, H. H. (1933): On *Mastacomys fuscus* Thomas. Transactions of the Royal Society of South Australia, vol. 57, pp. 125–9.

LANE, E. A. & RICHARDS, AOLA M. (1967): *Burramys parvus* Broom, a living fossil. Helictite, vol. 5, pp. 30–4.

213

MORCOMBE, M. K. (1967): The rediscovery after 83 years of the Dibbler *Antechinus apicalis* (Marsupialia, Dasyuridae). Western Australian Naturalist, vol. 10, pp. 102–11.

WEBSTER, H. O. (1962): Re-discovery of the Noisy Scrub-bird *Atrichornis clamosus*. Western Australian Naturalist, vol. 8, pp. 57–9, 81–4.

WILKINSON, E. H. (1961): The rediscovery of Leadbeater's Possum, *Gymnobelideus leadbeateri* McCoy. Victorian Naturalist, Melbourne, vol. 78, pp. 97–102.

WODZICKI, K. & FLUX, J. E. C. (1967): Re-discovery of the White-throated Wallaby, *Macropus parma* Waterhouse 1846, on Kawau Island, New Zealand. Australian Journal of Science, vol. 29, pp. 429–30.

The names of animals

CAIN, A. J. (1954): Animal species and their evolution. London: Hutchinson 190 pp.

HILL, J. E. (1966): The status of *Pipistrellus regulus* Thomas (Chiroptera, Vespertilionidae). Mammalia, Paris, vol. 30, pp. 302–7.

INTERNATIONAL CODE OF ZOOLOGICAL NOMENCLATURE (1964): London, 2 Edn., 176 pp.

The kinds of mammals

ANDERSON, S. & JONES, J. K. (Editors) (1967): Recent Mammals of the World. A Synopsis of Families. New York: Ronald Press. 453 pp.

JONES, R. (1968): The geographical background to the arrival of Man in Australia and Tasmania. Archaeology & Physical Anthropology in Oceania, vol. 3, pp. 186–215.

MERRILEES, D. (1968): Man the destroyer. Presidential Address. Journal of the Royal Society of Western Australia, vol. 51, pp. 1–24.

MORRIS, D. (1965): The Mammals. A Guide to the living species. London: Hodder and Stoughton. 448 pp.

RIDE, W. D. L. (1962): On the evolution of Australian marsupials. pp. 281–306 in The Evolution of Living Organisms. Ed. G. W. Leeper. Melbourne: Melbourne University Press, 459 pp.

RIDE, W. D. L. (1964): A review of Australian fossil marsupials. Presidential Address 1963. Journal of the Royal Society of Western Australia, vol. 47, pp. 97–131.

RIDE, W. D. L. (1968): On the past, present and future of Australian mammals. Presidential Address Section D: Zoology. Australian Journal of Science, vol. 31, pp. 1–11.

SIMPSON, G. G. (1953): Evolution and geography. (Condon Lecture). Eugene, Oregon: Oregon State System of Higher Education, 64 pp.

STORR, G. M. (1958): Are marsupials "second class" mammals? Western Australian Naturalist, vol. 6, pp. 179–83.

WALKER, E. P. (1964): Mammals of the World. Baltimore: Johns Hopkins Press. 3 vols, pp. 1–646, 647–1500, 1–769 (Bibliography).

Kangaroos and Wallabies

AUSTRALIAN CONSERVATION FOUNDATION (1967): Conservation of kangaroos. Australian Conservation Foundation Viewpoint Series no. 1. 6 pp.

BARKER, S. & BARKER, J. (1959): Physiology of the quokka. Journal of the Royal Society of Western Australia, vol. 42, pp. 72–6.

BERGER, PATRICIA J. (1966): Eleven-month 'embryonic diapause' in a marsupial. Nature, London, vol. 211, pp. 435–6.

BREEDEN, S. & BREEDEN, K. (1966): The life of the kangaroo. Melbourne: Angus & Robertson. 80 pp.

DEMPSTER, J. K. (1965): Poto-Roo. Wildlife, vol. 2, Pt 4, pp. 18–22.

FINLAYSON, H. H. (1927): Observations on the South Australian members of the subgenus 'Wallabia'. Transactions of the Royal Society of South Australia, vol. 51, pp. 363–77.

FINLAYSON, H. H. (1930): Observations on the South Australian species of the subgenus 'Wallabia'. Part II. Transactions of the Royal Society of South Australia, vol. 54, pp. 47–56.

FINLAYSON, H. H. (1931): On mammals from the Dawson Valley, Queensland. Part I. Transactions of the Royal Society of South Australia, vol. 55, pp. 67–89.

FRITH, H. J. & CALABY, J. H. (1968): Kangaroos. Melbourne: F. W. Cheshire.

GRIFFITHS, M. & BARKER, R. (1966): The plants eaten by sheep and by kangaroos grazing together in a paddock in south-western Queensland. C.S.I.R.O. Wildlife Research, vol. 11, pp. 145–67.

GUILER, E. R. (1957): Longevity in the wild potoroo: *Potorous tridactylus* (Kerr). Australian Journal of Science, vol. 20, p. 26.

GUILER, E. R. (1958): Observations on a population of small marsupials in Tasmania. Journal of Mammalogy, vol. 39, pp. 44–58.

GUILER, E. R. (1960): The breeding season of *Potorous tridactylus* (Kerr). Australian Journal of Science, vol. 23, pp. 126–7.

GUILER, E. R. (1960): The pouch young of the potoroo. Journal of Mammalogy, vol. 41, pp. 441–51.

GUILER, E. R. & KITCHENER, D. A. (1967): Further observations on longevity in the wild potoroo, *Potorous tridactylus*. Australian Journal of Science, vol. 30, pp. 105–6.

HARRISON, J. L. (1962): Mammals of Innisfail. I. Species and distribution. Australian Journal of Zoology, vol. 10, pp. 45–83.

HARRISON, J. L. (1962): The food of some Innisfail mammals. Proceedings of the Royal Society of Queensland, vol. 73, pp. 37–43.

HEINSOHN, G. E. (1968): Habitat requirements and reproductive potential of the macropod marsupial *Potorous tridactylus* in Tasmania. Mammalia, Paris, vol. 32, pp. 30–43.

HUGHES, R. D. (1965): On the age composition of a small sample of individuals from a population of the Banded Hare Wallaby, *Lagostrophus fasciatus* (Peron & Lesueur). Australian Journal of Zoology, vol. 13, pp. 75–95.

MOLLISON, B. (1960): Progress report on the ecology and control of marsupials in the Florentine Valley. Appita, vol. 14, no. 2, pp. 21–7.

PACKER, W. C. (1963): Homing behaviour in the Quokka, *Setonix brachyurus* (Quoy and Gaimard) (Marsupialia). Journal of the Royal Society of Western Australia, vol. 46, pp. 28–32.

RIDE, W. D. L., MEES, G. F., DOUGLAS, A. M., ROYCE, R. D., & TYNDALE-BISCOE, C. H. (1962): The results of an expedition to Bernier and Dorre Islands, Shark Bay, Western Australia in July 1959. Fauna Bulletin Western Australia, no. 2, pp. 1–131.

ROFF, C. & KIRKPATRICK, T. (1962): The kangaroo industry in Queensland. Queensland Journal of Agricultural Science, vol. 19, pp. 385–401.

ROFF, C. & KIRKPATRICK, T. (1966): The kangaroo industry in Queensland. First supplement, 1961–1965. Queensland Journal of Agricultural and Animal Sciences, vol. 23, pp. 467–73.

SHARMAN, G. B. (1955): Studies on marsupial reproduction. III. Normal and delayed pregnancy in *Setonix brachyurus*. Australian Journal of Zoology, vol. 3, pp. 56–70.

SHIELD, J. W. (1959): Rottnest field studies concerned with the quokka. Journal of the Royal Society of Western Australia, vol. 42, pp. 76–8.

STODART, E. (1966): Observations on the behaviour of the marsupial *Bettongia lesueuri* (Quoy & Gaimard) in an enclosure. C.S.I.R.O. Wildlife Research, vol. 11, pp. 91–101.

STORR, G. M. (1963): Some factors inducing change in the vegetation of Rottnest Island. Western Australian Naturalist, vol. 9, pp. 15–22.

TYNDALE-BISCOE, C. H. (1968): Reproduction and post-natal development in the marsupial *Bettongia lesueur* (Quoy and Gaimard). Australian Journal of Zoology, vol. 16, pp. 577–602.

WAKEFIELD, N. A. (1961): Victoria's Rock-wallabies. Victorian Naturalist, Melbourne, vol. 77, pp. 322–32.

WAKEFIELD, N. A. (1963): Notes on Rock-wallabies. Victorian Naturalist, Melbourne. vol. 80, pp. 169–76.

Possums, the Koala and wombats

BOLLIGER, A. (1962): Gravel in the caecum of the koala (*Phascolarctos cinereus*). Australian Journal of Science, vol. 24, p. 416.

BOLLIGER, A. & FINCKH, E. S. (1962): The prevalence of cryptococcosis in the koala (*Phascolarctos cinereus*). Medical Journal of Australia, 1962, pp. 545–7.

BRAZENOR, C. W. (1932): A re-examination of *Gymnobelideus leadbeateri* McCoy. Australian Zoologist, vol. 7, pp. 106–9.

BRAZENOR, C. W. (1946): Last chapter to come. A history of Victoria's rarest possum. Wild Life, Melbourne, vol. 8, pp. 383–5.

CALABY, J. H. (1957): A new record of the Scaly-tailed Possum (*Wyulda squamicaudata* Alexander). Western Australian Naturalist, vol. 5, pp. 186–91.

CLARK, MEREDITH J. (1967): Pregnancy in the lactating pigmy possum *Cercartetus concinnus*. Australian Journal of Zoology, vol. 15, pp. 673–83.

DAHL, K. (1897): *Pseudochirus dahlii*, 'Wogoit'. pp. 203–5 in Biological notes on north-Australian Mammalia. The Zoologist, Series 4, vol. 1, pp. 189–216.

FINLAYSON, H. H. (1961): Mitchell's Wombat in South Australia. Transactions of the Royal Society of South Australia, vol. 85, pp. 207–15.

FLEAY, D. (1932): The Pigmy Flying Possum. Victorian Naturalist, Melbourne, vol. 49, pp. 165–71.

FLEAY, D. (1941): Queensland's rare striped possum is dainty, skilful, a fighter—and odorous. Australasian, vol. 150, no. 4825, June 21, p. 29.

FLEAY, D. (1947): Gliders of the gum trees. Melbourne: Bread and Cheese Club. 113 pp.

FLEAY, D. (1963): Trapeze artist of the tall timber. Wildlife in Australia, vol. 1, no. 2, pp. 3–6.

GUILER, ERIC R. & BANKS, DORIS M. (1958): A further examination of the distribution of the Brush Possum, *Trichosurus vulpecula*, in Tasmania. Ecology, vol. 39, pp. 89–97.

HARRISON, E. (1961): Tuans and Gliders. Victorian Naturalist, Melbourne, vol. 78, pp. 224–31.

HUGHES, R. L., THOMSON, J. A. & OWEN, W. H. (1965): Reproduction in natural populations of the Australian ringtail possum, *Pseudocheirus peregrinus* (Marsupialia: Phalangeridae), in Victoria. Australian Journal of Zoology, vol. 13, pp. 383–406.

LANE, E. A. & RICHARDS, AOLA M. (1967): *Burramys parvus* Broom, a living fossil. Helictite, vol. 5, pp. 30–4.

LYNE, A. G. & VERHAGEN, A. M. W. (1957): Growth of the marsupial *Trichosurus vulpecula* and a comparison with some higher mammals. Growth, vol. 21, pp. 167–95.

MC NALLY, J. (1955): Damage to Victorian exotic pine plantations by native animals. Australian Forestry, vol. 19, pp. 87–99.

MC NALLY, J. (1957): A field survey of a Koala population. Proceedings of the Royal Zoological Society of New South Wales 1955–56, pp. 18–27.

MC NALLY, J. (1960): Koala management in Victoria. Australian Museum Magazine, vol. 13, pp. 178–81.

MARLOW, B. J. (1965): Wombats. Australian Natural History, vol. 15, pp. 65–9.

MARSH, MICHAEL (1967): Ring-tailed possums. Australian Natural History, vol. 15, pp. 294–7.

MOLLISON, B. (1960): Progress report on the ecology and control of marsupials in the Florentine Valley. Appita, vol. 14, part 2, pp. 21–7.

NICHOLSON, P. J. (1963): Wombats. Timbertop Magazine, Geelong Grammar School, no. 8, pp. 32–8.

PILTON, P. E. & SHARMAN, G. B. (1962): Reproduction in the marsupial *Trichosurus vulpecula*. Journal of Endocrinology, vol. 25, pp. 119–36.

PRATT, AMBROSE (1937): The call of the Koala. Melbourne: Robertson & Mullens. 120 pp.

RAND, A. L. (1937): Results of the Archbold Expeditions No. 17. Some original observations on the habits of *Dactylopsila trivirgata* Gray. American Museum Novitates, no. 957, 7 pp.

RIDPATH, P. (1967): Possum moods. Sydney: Ure Smith. 72 pp.

RYAN, MARK (1963): Feeding habits of Leadbeater's possum. Victorian Naturalist, Melbourne, vol. 79, p. 275.

THOMSON, J. A. & OWEN, W. H. (1963): The ringtail possum in south-eastern Australia. Australian Natural History, vol. 14, pp. 145–8.

THOMSON, J. A. & OWEN, W. H. (1964): A field study of the Australian ringtail possum *Pseudocheirus peregrinus* (Marsupialia: Phalangeridae). Ecological Monographs, vol. 34, pp. 27–52.

THOMSON, J. A. & PEARS, F. N. (1962): The functions of the anal glands of the Brushtail Possum. Victorian Naturalist, Melbourne, vol. 78, pp. 306–8.

TYNDALE-BISCOE, C. H. (1955): Observations on the reproduction and ecology of the Brush-tailed Possum, *Trichosurus vulpecula* Kerr (Marsupialia), in New Zealand. Australian Journal of Zoology, vol. 3, pp. 162–84.

VAN DEUSEN, H. M. & STEARNS, E. I. (1961): Source of color in the fur of the Green Ring-tailed Possum. Journal of Mammalogy, vol. 42, pp. 149–52.

WAKEFIELD, N. A. (1961): Notes on the Tuan. Victorian Naturalist, Melbourne, vol. 78, pp. 232–5.

WAKEFIELD, N. A. (1963): The Australian pigmy-possums. Victorian Naturalist, Melbourne, vol. 80, pp. 99–116.

WILKINSON, E. H. (1961): The rediscovery of Leadbeater's Possum, *Gymnobelideus leadbeateri* McCoy. Victorian Naturalist, Melbourne, vol. 78, pp. 97–102.

Bandicoots

GUILER, E. R. (1958): Observations on a population of small marsupials in Tasmania. Journal of Mammalogy, vol. 39, pp. 44–58.

HEINSOHN, GEORGE E. (1966): Ecology and reproduction of the Tasmanian bandicoots (*Perameles gunni* and *Isoodon obesulus*). University of California publications in Zoology, vol. 80, pp. 1–96.

KREFFT, G. (1866): *Chaeropus occidentalis*. pp. 12–14 in On the vertebrated animals of the Lower Murray and Darling, their habits, economy, and geographical distribution. Transactions of the Philosophical Society of N.S.W., 1862–1865, pp. 1–33.

LYNE, A. G. (1964): Australian bandicoots. Australian Natural History, vol. 14, pp. 281–5.

LYNE, A. G. (1964): Observations on the breeding and growth of the marsupial *Perameles nasuta* Geoffroy, with notes on other bandicoots. Australian Journal of Zoology, vol. 12, pp. 322–39.

SMYTH, D. R. & PHILPOTT, C. M. (1968): A field study of the Rabbit Bandicoot, *Macrotis lagotis*, Marsupialia, from central Western Australia. Transactions of the Royal Society of South Australia, vol. 92, pp. 3–14.

STODART, E. (1966): Management and behaviour of breeding groups of the marsupial *Perameles nasuta* Geoffroy in captivity. Australian Journal of Zoology, vol. 14, pp. 611–23.

Native-cats and their relatives

CALABY, J. H. (1960): The Numbat of south-western Australia. Australian Museum Magazine, vol. 13, pp. 143–6.

CALABY, J. H. (1960): Observations on the Banded Anteater *Myrmecobius f. fasciatus* Waterhouse (Marsupialia), with particular reference to its food habits. Proceedings of the Zoological Society of London, vol. 135, pp. 183–207.

FLEAY, D. (1952): The Tasmanian or marsupial Devil—its habits and family life. Australian Museum Magazine, vol. 10, pp. 275–80.

FLEAY, D. (1965): Australia's 'needle-in-a-haystack' marsupial. Vicissitudes in the pursuit and study of Ingram's Planigale, the smallest pouch-bearer. Victorian Naturalist, Melbourne, vol. 82, pp. 195–204.

FLEAY, D. (1963): Have you met the Northern Quoll? Wildlife in Australia, vol. 1, no. 1, pp. 27–9.

FLEAY, D. (1965): Breeding the Mulgara. Wildlife in Australia, vol. 3, no. 1, pp. 2–5.

GREEN, R. H. (1967): Notes on the Devil (*Sarcophilus harrisii*) and the Quoll (*Dasyurus viverrinus*) in north-eastern Tasmania. Records of the Queen Victoria Museum, Launceston, no. 27, pp. 1–13.

GUILER, E. R. (1961): The former distribution and decline of the Thylacine. Australian Journal of Science, vol. 23, pp. 207–10.

GUILER, E. R. (1964): Tasmanian Devils. Australian Natural History, vol. 14, pp. 360–2.

GUILER, E. R. & MELDRUM, G. K. (1958): Suspected sheep killing by the Thylacine, *Thylacinus cynocephalus*. Australian Journal of Science, vol. 20, pp. 214–5.

HARRISON, E. (1961): Tuans and Gliders. Victorian Naturalist, Melbourne, vol. 78, pp. 224–31.

HORNER, B. ELIZABETH & TAYLOR, J. MARY (1959): Results of the Archbold Expeditions. No. 80. Observations on the biology of the Yellow-footed Marsupial Mouse, *Antechinus flavipes flavipes*. American Museum Novitates, no. 1972, pp. 1–24.

JONES, F. WOOD (1923): Marsupial Mole, or Pouched Mole, pp. 128–31 in The Mammals of South Australia Pt. 1. The Monotremes and the carnivorous marsupials. Adelaide: Govt. Printer, 131 pp. (Reprinted 1969).

MARLOW, B. J. (1961): Reproductive behaviour of the marsupial mouse, *Antechinus flavipes* (Waterhouse) (Marsupialia) and the development of the pouch young. Australian Journal of Zoology, vol. 9, pp. 203–18.

MORCOMBE, MICHAEL K. (1967): Discovery of a long-lost Australian. Australian Women's Weekly, vol. 34, no. 52, pp. 18–21.

RIDE, W. D. L. (1965): Locomotion in the Australian marsupial *Antechinomys*. Nature, London, vol. 205, p. 199.

SCHMIDT-NIELSEN, K. & NEWSOME, A. E. (1962): Water balance in the Mulgara (*Dasycercus cristicauda*), a carnivorous desert marsupial. Australian Journal of Biological Science, vol. 15, pp. 683–9.

WAKEFIELD, N. A. (1961): Honey for phalangers and phascogales. Victorian Naturalist, Melbourne, vol. 78, pp. 192–9.

WAKEFIELD, N. A. & WARNEKE, R. M. (1963): Some revision in *Antechinus* (Marsupialia)— 1. Victorian Naturalist, Melbourne, vol. 80, pp. 194–219.

WAKEFIELD, N. A. & WARNEKE, R. M. (1967): Some revision in *Antechinus* (Marsupialia)— 2. Victorian Naturalist, Melbourne, vol. 84, pp. 69–99.

WOOLLEY, P. (1966): Reproduction in *Antechinus* spp. and other dasyurid marsupials. Symposia of the Zoological Society of London, no. 15, pp. 281–94.

Australian native rats and mice

BARROW, G. J. (1964): *Hydromys chrysogaster*—some observations. Queensland Naturalist, vol. 17, pp. 43–4.

CALABY, J. H. & WIMBUSH, D. J. (1964): Observations on the Broad-toothed Rat, *Mastacomys fuscus* Thomas. C.S.I.R.O. Wildlife Research, vol. 9, pp. 123–33.

FINLAYSON, H. H. (1939): On mammals from the Lake Eyre Basin. Pt V. General remarks on the increase of murids and their population movements in the Lake Eyre Basin during the years 1930–36. Transactions of the Royal Society of South Australia, vol. 63, pp. 348–53.

FINLAYSON, H. H. (1961): A re-examination of *Mesembriomys hirsutus* Gould 1842 (Muridae). Transactions of the Royal Society of South Australia, vol. 84, pp. 149–62.

FLEAY, D. (1964): The rat that mastered the waterways. Wildlife in Australia, vol. 1, no. 4, pp. 3–7.

GREEN, R. H. (1967): The murids and small dasyurids in Tasmania. Records of the Queen Victoria Musuem, Launceston, no. 28, 19 pp.

HORNER, B. ELIZABETH & TAYLOR, J. MARY (1965): Systematic relationships among *Rattus* in southern Australia: evidence from cross-breeding experiments. C.S.I.R.O. Wildlife Research, vol. 10, pp. 101–9.

HARRISON, J. L. (1962): Mammals of Innisfail. I. Species and distribution. Australian Journal of Zoology, vol. 10, pp. 45–83.

HARRISON, J. L. (1962): The food of some Innisfail mammals. Proceedings of the Royal Society of Queensland, vol. 73, pp. 37–43.

JONES, F. WOOD (1925): *Leporillus* (Thomas, 1906), pp. 326–36 in Mammals of South Australia Pt 3. The Monodelphia. Adelaide: Government Printer, pp. 271–458. (Reprinted 1969).

MC DOUGALL, W. A. (1944): An investigation of the rat pest problem in Queensland cane fields: 2. Species and general habits. Queensland Journal of Agricultural Science, vol. 1, no. 2, pp. 48–78.

MC DOUGALL, W. A. (1946): An investigation of the rat pest problem in Queensland cane fields: 4. Breeding and life histories. Queensland Journal of Agricultural Science, vol. 3, pp. 1–43.

MAC MILLEN, R. E. & LEE, A. K. (1967): Australian desert mice: Independence of exogenous water. Science, vol. 158, no. 3799, pp. 383–5.

MC NALLY, J. (1955): Damage to Victorian exotic pine plantations by native animals. Australian Forestry, vol. 19, pp. 87–99.

MC NALLY, J. (1960): The biology of the water rat *Hydromys chrysogaster* Geoffroy (Muridae: Hydromyinae) in Victoria. Australian Journal of Zoology, vol. 8, pp. 170–80.

RIDE, W. D. L. & TYNDALE-BISCOE, C. H. (1962): The Ashy-grey mouse, pp. 71–7 in Mammals. Western Australian Fauna Bulletin no. 2, pp. 54–97.

TAYLOR, J. MARY (1961): Reproductive biology of the Australian bush rat, *Rattus assimilis*. University of California Publications in Zoology, vol. 60, pp. 1–66.

TROUGHTON, E. LE G. (1923): A revision of the rats of the genus *Leporillus* and the status of *Hapalotis personata* Krefft. Records of the Australian Museum, vol. 14, pp. 23–41.

WARNEKE, R. M. (1960): Rediscovery of the Broad-toothed Rat in Gippsland. Victorian Naturalist, Melbourne, vol. 77, pp. 195–6.

Australian bats

DOUGLAS, A. M. (1967): The natural history of the Ghost Bat, *Macroderma gigas* (Microchiroptera, Megadermatidae), in Western Australia. Western Australian Naturalist, vol. 10, pp. 125–38.

DWYER, P. D. (1963): Bat banding. Australian Natural History, vol. 14, pp. 198–200.

DWYER, P. D. (1968): The biology, origin, and adaptation of *Miniopterus australis* (Chiroptera) in New South Wales. Australian Journal of Zoology, vol. 16, pp. 49–68.

GEORGE, G. & WAKEFIELD, N. A. (1961): Victorian cave bats. Victorian Naturalist, Melbourne, vol. 77, pp. 294–302.

GREEN, R. H. (1965): Observations on the little brown bat, *Eptesicus pumilus* Gray in Tasmania. Records of the Queen Victoria Museum, Launceston, no. 20, pp. 1–16.

GREEN, R. H. (1966): Notes on Lesser Long-eared Bat *Nyctophilus geoffroyi* in northern Tasmania. Records of the Queen Victoria Museum, Launceston, no. 22, pp. 1–4.

HAMILTON-SMITH, E. (1964): Swimming of bats. Victorian Naturalist, Melbourne, vol. 81, p. 104.

HAMILTON-SMITH, E. (1966): The geographical distribution of Australian cave-dwelling Chiroptera. International Journal of Speleology, vol. 2, pp. 91–104.

MC KEAN, JOHN L. & HALL, L. S. (1965): Distribution of the Large-footed Myotis, *Myotis adversus*, in Australia. Victorian Naturalist, Melbourne, vol. 82, pp. 164–8.

MC KEAN, JOHN L. & HAMILTON-SMITH, ELERY (1967): Litter size and maternity sites in Australian bats (Chiroptera). Victorian Naturalist, Melbourne, vol. 84, pp. 203–6.

MARLOW, B. J. (1961): Vampire bats—true and false. Australian Museum Magazine, vol. 13, pp. 354–7.

NELSON, J. E. (1962): Flying foxes. Australian Natural History, vol. 14, pp. 12–14.

NELSON, J. E. (1964): Notes on *Syconycteris australis*, Peters, 1867 (Megachiroptera). Mammalia, Paris, vol. 28, pp. 429–32.

NELSON, J. E. (1964): Vocal communication in Australian flying foxes (Pteropodidae; Megachiroptera). Zeitschrift für Tierpsychologie, vol. 21, pp. 857–70.

NELSON, J. E. (1965): Behaviour of Australian Pteropodidae (Megachiroptera). Animal Behaviour, vol. 13, pp. 544–57.

NELSON, J. E. (1965): Movements of Australian flying foxes (Pteropodidae: Megachiroptera). Australian Journal of Zoology, vol. 13, pp. 53–73.

PURCHASE, D. (1962): A second report on bat-banding in Australia. Technical Paper, Division of Wildlife Research C.S.I.R.O. no. 2, pp. 1–16.

PURCHASE, D. & HISCOX, P. M. (1960): A first report on bat-banding in Australia. C.S.I.R.O. Wildlife Research, vol. 5, pp. 44–51.

RATCLIFFE, F. N. (1931): The flying fox (*Pteropus*) in Australia. Bulletin, Council for Scientific and Industrial Research, Melbourne, no. 53, 81 pp.

RATCLIFFE, F. N. (1938): Flying Fox and Drifting Sand. London: Chatto & Windus. Reprinted in Australian Edn in 1947 and subsequently. Sydney: Angus & Robertson. 332 pp.

SEEBECK, J. H. & HAMILTON-SMITH, E. (1967): Notes on a wintering colony of bats. Victorian Naturalist, Melbourne, vol. 84, pp. 348–51.

SIMPSON, D. G. & HAMILTON-SMITH, E. (1965): Third and fourth annual reports on bat-banding in Australia. Technical Paper, Division of Wildlife Research C.S.I.R.O. no. 9, pp. 1–24.

WILSON, S. J., LANE, S. G. & MCKEAN, J. L. (1965): The use of mist nets in Australia. Technical Paper, Division of Wildlife Research C.S.I.R.O. no. 8, pp. 2–36.

Non-marsupial carnivores: The Dingo and seals

CARRICK, R. & INGHAM, S. E. (1962): Studies on the southern elephant seal *Mirounga leonina* (L.). C.S.I.R.O. Wildlife Research, vol. 7, pp. 89–206.

CSORDAS, S. E. & INGHAM, SUSAN E. (1965): The New Zealand fur seal, *Arctocephalus forsteri* (Lesson), at Macquarie Island, 1949–64. C.S.I.R.O. Wildlife Research, vol. 10, pp. 83–99.

INGHAM, S. E. (1960): The status of seals (Pinnipedia) at Australian Antarctic stations. Mammalia, Paris, vol. 24, pp. 422–30.

KING, JUDITH E. (1964): Seals of the world. London: Trustees of British Museum (Natural History). 154 pp.

MACINTOSH, N. W. G. (1956): Trail of the dingo. Etruscan, Staff magazine of the Bank of N.S.W., vol. 5, no. 4, pp. 8–12.

MARLOW, B. J. (1962): Dingoes. Australian Natural History, vol. 14, pp. 61–4.

TOMLINSON, A. R. (1955): Wild dogs and dingoes in Western Australia. Journal of Agriculture, Western Australia, series 3, vol. 4, pp. 4–12.

WARNEKE, R. M. (1966): Seals of Westernport. Victoria's Resources, vol. 8, no. 2, pp. 2–4

The Monotremes

BARRETT, C. (1944): The Platypus. Melbourne: Robertson & Mullens. 62 pp.

BARRETT, C. (1947): The Platypus. Natural History, New York, vol. 56, p. 342.

BURRELL, H. (1927): The Platypus: its discovery, zoological position, form and characteristics, habits, life history, etc. Sydney: Angus & Robertson. 227 pp.

CALABY, J. H. (1968): The Platypus (*Ornithorhynchus anatinus*) and its venomous characteristics, Chapter I, pp. 15–29 in Venomous Animals and their venoms, vol. 1 Ed. W. Bücherl, E. Buckley and V. Deulofeu. New York: Academic Press.

COLEMAN, E. (1934): The Echidna under domestication. Victorian Naturalist, Melbourne, vol. 51, pp. 12–21.

COLEMAN, E. (1935): The Echidna under domestication. Victorian Naturalist, Melbourne, vol. 52, pp. 151–4.

COLEMAN, E. (1935): Hibernation and other habits of the Echidna under domestication. Victorian Naturalist, Melbourne, vol. 52, pp. 55–61.

COLEMAN, E. (1938): Notes on hibernation, ecdysis, and sense of smell of the Echidna under domestication. Victorian Naturalist, Melbourne, vol. 55, pp. 105–7.

FLEAY, D. (1944): We breed the Platypus. Melbourne: Robertson & Mullens. 44 pp.

GRIFFITHS, M. (1965): Rate of growth and intake of milk in a suckling Echidna. Comparative Biochemistry and Physiology, vol. 16, pp. 383–92.

GRIFFITHS, M. & SIMPSON, K. G. (1966): A seasonal feeding habit of spiny ant-eaters. C.S.I.R.O. Wildlife Research, vol. 11, pp. 137–43.

The rare ones and those with changed status

CALABY, J. H. (1966): Mammals of the Upper Richmond and Clarence Rivers, New South Wales. Technical Paper, Division of Wildlife Research C.S.I.R.O. no. 10, pp. 1–55.

FINLAYSON, H. H. (1961): On Central Australian mammals. IV. The distribution and status of Central Australian species. Records of the South Australian Museum, vol. 14, pp. 141–91.

HARRISON, J. L. (1962): Mammals of Innisfail. I. Species and distribution. Australian Journal of Zoology, vol. 10, pp. 45–83.

JOHNSON, D. H. (1964): Mammals, in Records of the American-Australian Scientific Expedition to Arnhem Land, vol. 4, Zoology, pp. 427–515. Ed. R. L. Specht, Melbourne: Melbourne University Press.

KIRKPATRICK, T. H. (1966): Mammals, birds and reptiles of the Warwick District, Queensland. 1. Introduction and mammals. Queensland Journal of Agricultural and Animal Sciences, vol. 23, pp. 591–8.

MARLOW, B. (1958): A survey of the marsupials of New South Wales. C.S.I.R.O. Wildlife Research, vol. 3, pp. 71–114.

RIDE, W. D. L. (1964): General narrative and the mammals of Depuch Island. Special Publications Western Australian Museum, no. 2, pp. 13–22, 75–8.

RIDE, W. D. L., MEES, G. F., DOUGLAS, A. M., ROYCE, R. D., & TYNDALE-BISCOE, C. H. (1962): The results of an expedition to Bernier and Dorre Islands, Shark Bay, Western Australia in July 1959. Fauna Bulletin Western Australia, no. 2, pp. 1–131.

SERVENTY, V. N. (1953): Mammals. Pt 4 of Australian Geographical Society Reports no. 1. The Archipelago of the Recherche, pp. 40–8.

WAKEFIELD, N. A. (1966): Mammals of the Blandowski Expedition to North-western Victoria. Proceedings of the Royal Society of Victoria, vol. 79, pp. 371–91.

WAKEFIELD, N. A. (1966): Mammals recorded for the Mallee, Victoria. Proceedings of the Royal Society of Victoria, vol. 79, pp. 627–36.

The how and why of conservation

BAUR, G. N. (1964): The Future of Rainforest in New South Wales. Forestry Commission of N.S.W. Technical Paper, no. 5, 12 pp.

BROWNE, W. R., COSTIN, A. B., DAY, M. F., & TURNER, J. S. (1965): The Kosciusko Primitive Area. Australian Journal of Science, vol. 27, pp. 203–7.

CROWCROFT, P. (1964): Nature Conservation in South Australia. Proceedings of the Royal Geographical Society of Australasia. South Australian Branch. vol. 65, pp. 31–41.

DAY, M. F. C. et al. (1968): National Parks and Reserves in Australia. [Report of the Committee established by Council in August 1958]. Australian Academy of Science, pp. i–iii, 1–45.

DEMPSTER, J. K. (1961): Kangaroos Management in Victorian agricultural areas. Journal of Agriculture, Victorian Department of Agriculture, vol. 59, pp. 45–52.

HEMINGWAY, D. (1963): The 'new look' in wildlife conservation. Victoria's Resources, vol. 5, no. 1, pp. 21–3.

KING, A. R. (1963): Report on the influence of colonization on the forests and the prevalence of bushfires in Australia. C.S.I.R.O., Division of Physical Chemistry, 72 pp.

NEWLAND, B. C. (1963): May sheep safely graze?—A proposal for reserves in pastoral areas. Presidential Address. Proceedings of the Royal Geographical Society of Australasia. South Australian Branch. vol. 65, pp. 43–51.

SPECHT, R. L. (with collaboration from J. B. Cleland) (1961, 1963): Flora Conservation in South Australia. Transactions of the Royal Society of South Australia, vol. 85, pp. 177–96; vol. 87, pp. 63–92.

WARNEKE, R. M. (1963): Rare today—tomorrow . . . ? Victoria's Resources, vol. 5, no. 1, pp. 24–5.

WEBB, L. J. (1966): The identification and conservation of habitat-types in the wet tropical lowlands of North Queensland. Proceedings of the Royal Society of Queensland, vol. 78, pp. 59–86.

Appendix II For the student and professional user

This book will come into use several years before the publication of our descriptive work on Australian mammals which is to be published by the Clarendon Press; it will also appear well in advance of the *Checklist of Australian Mammals* by J. H. Calaby, J. A. Mahoney, and W. D. L. Ride upon which its taxonomy is based.

So that it will be of greater use to the student who wishes to identify specimens with more precision than is possible from the external characters given here, this appendix is included. It consists of a list of the species in which each is accompanied by one or more bibliographical references to morphological descriptions of both external features and skulls and teeth. The descriptions which are cited differ widely in quality but are as representative as I can find of the species concepts used in this work; in using them, it should be borne in mind that the authors possessed varying ideas of the amount of permissible variation within the species and, in many examples, the precise descriptions are rather narrow in their limits. Where a specimen lies beyond a definition given, a student may have to consult the description of closely similar species in the list as well and decide which is most applicable.

References to the titles of journals are abbreviated in conformity with the *World List of Scientific Periodicals*. They are given in alphabetical order of their authorship at the end of the taxonomic list.

MARSUPIALIA

MACROPODIDAE:

*Macropus giganteus** Shaw, 1790. Thomas (1888) pp. 15–19 [including both *M. giganteus*, var. *typicus* and *M. giganteus*, var. *fuliginosus* with such deletions as are made necessary by Kirsch, J. A. W. & Poole, W. E. (1967) *Nature*, vol. 215, pp. 1097–8].

*M. fuliginosus** (Desmarest, 1817). Thomas (1888) p. 20 [as *M. giganteus*, var. *melanops*]. Jones (1923–5) pp. 255–261 [including the morphological range represented by 'The typical South Australian animal' and *M. fuliginosus*].

M. robustus Gould, 1941. Thomas (1888) pp. 23–25 [including *M. robustus* and *M. isabellinus*]. Waite (1901) pp. 131–134 [as *M. isabellinus*]. Jones (1923–5) pp. 250–253.

M. antilopinus (Gould, 1842). Thomas (1888) pp. 21–22. Johnson (1964) pp. 460–464 [as *Osphranter antilopinus*].

Megaleia rufa (Desmarest, 1822). Thomas (1888) pp. 25–27. Jones (1923–5) pp. 253–255.

Macropus agilis (Gould, 1842). Thomas (1888) pp. 42–43.

M. rufogriseus (Desmarest, 1817). Thomas (1888) pp. 32–35 [including both *M. ruficollis* var. *typicus* and var. *bennetti*].

M. dorsalis (Gray, 1837). Thomas (1888) pp. 37–38. Finlayson (1931) pp. 72–75.

M. parryi (Bennett, 1835). Thomas (1888) pp. 39–40. Finlayson (1931) pp. 75–78.

---

* Descriptions of these two forms in the existing literature are almost invariably confused and contain elements of both species. Until this is corrected, in most of Australia location is the best guide to identification.

M. irma (Jourdan, 1837). Thomas (1888) pp. 40–41.

M. greyi (Waterhouse, 1846). Finlayson (1927) pp. 363–377.

M. bernardus (Rothschild, 1904). Rothschild (1904) p. 414 [as *Dendrodorcopsis woodwardi*]. Le Souef & Burrell (1926) p. 186.

M. eugenii (Desmarest, 1817). Thomas (1888) pp. 54–57. Jones (1923–5) pp. 235–242 [including *Thylogale eugenii* and *T. flindersi*].

M. parma Waterhouse, 1946. Ride (1957) pp. 327–346 [as *Protemnoden parma*].

Wallabia bicolor (Desmarest, 1804). Finlayson (1931) pp. 79–81 [as *Macropus* (*W.*) *ualabatus*, var. *ingrami*].

Thylogale billardierii (Desmarest, 1822). Thomas (1888) pp. 58–60 [as *Macropus billardieri*].

T. thetis (Lesson, 1827). Thomas (1888) pp. 52–54 [as *Macropus thetidis*].

T. stigmatica Gould, 1860. Thomas (1888) pp. 44–49 [including *M. coxeni*, *M. stigmaticus* and *M. wilcoxi*].

Setonix brachyurus (Quoy & Gaimard, 1830). Thomas (1888) pp. 60–62 [as *Macropus brachyurus*].

Onychogalea unguifera (Gould, 1841). Thomas (1888) pp. 74–75.

O. lunata (Gould, 1840). Thomas (1888) pp. 77–79.

O. fraenata (Gould, 1841). Thomas (1888) pp. 75–77.

Lagorchestes conspicillatus Gould, 1842. Thomas (1888) pp. 80–82.

L. hirsutus (Gould, 1844). Thomas (1888) pp. 84–86.

L. leporides (Gould, 1841). Thomas (1888) pp. 82–84.

L. asomatus Finlayson, 1943. Finlayson (1943a) pp. 319–321.

Lagostrophus fasciatus (Péron, 1807). Thomas (1887) pp. 544–547. Thomas (1888) pp. 100–102.

Petrogale penicillata (Griffith, 1827). Thomas (1888) pp. 66–71 [including *P. penicillata*, *P. lateralis*, *P. inornata*]. Finlayson (1931) pp. 82–85 [*P.p.* var. *herberti*]. Jones (1923–5) pp. 227–232 [including *P. lateralis* and *P. pearsoni*].

P. rothschildi Thomas, 1904. Le Souef & Burrell (1926) p. 207.

P. purpureicollis Le Souef, 1924. Le Souef & Burrell (1926) p. 208.

P. godmani Thomas, 1923. Thomas (1923) pp. 177–178.

P. brachyotis Gould, 1841. Thomas (1888) pp. 69–70.

P. xanthopus Gray, 1855. Thomas (1888) pp. 65–66.

Peradorcas concinna (Gould, 1842). Tate (1948) pp. 280–285.

Dendrolagus lumholtzi Collett, 1844. Rothschild & Dollman (1936) pp. 501–502, 534, 548.

D. bennettianus De Vis, 1887. Rothschild & Dollman (1936) pp. 499–501, 532, 548.

Hypsiprymnodon moschatus Ramsay, 1876. Thomas (1888) pp. 122–124. Carlsson (1915) pp. 3–48.

Caloprymnus campestris (Gould, 1843). Finlayson (1932) pp. 148–167.

Aepyprymnus rufescens (Gray, 1837). Finlayson (1931) pp. 85–89.

Bettongia penicillata Gray, 1837. Finlayson (1958) pp. 268–290 [subspecies excluding *B. p. anhydra*]. Wakefield (1967) pp. 8–22.

B. gaimardi (Desmarest, 1822). Finlayson (1959) pp. 283–289 [as *B. cuniculus*]. Wakefield (1967) pp. 8–22.

B. tropica Wakefield, 1967. Wakefield (1967) pp. 8–22.

B. lesueur (Quoy & Gaimard, 1824). Finlayson (1958) pp. 238–268.

Potorous tridactylus (Kerr, 1792). Thomas (1888) pp. 117–121 [only that part of the description which applies to "the smaller New South Wales examples" together with *P. gilberti*].

P. apicalis (Gould, 1851). Thomas (1888) pp. 117–120 [only that part of the description which applies to "the large Tasmanian form ("*apicalis*")" and the "still smaller Tasmanian form described as "*rufus*" "].

P. platyops (Gould, 1844). Thomas (1888) pp. 121–122. Finlayson (1938) pp. 132–140 [as *Potorous morgani*].

PHALANGERIDAE:

Trichosurus vulpecula (Kerr, 1792). Thomas (1888) pp. 187–191. Lyne & Verhagen (1957) pp. 167–195.

T. arnhemensis Collett, 1897. Collett (1897) pp. 328–329 [under *Trichosurus vulpecula*]. Johnson (1964) pp. 450–451.

T. caninus (Ogilby, 1836). Thomas (1888) pp. 191–192.

Wyulda squamicaudata Alexander, 1919. Finlayson (1943) pp. 255–261. Calaby (1957) pp. 186–191.

Phalanger maculatus (Desmarest, 1818). Thomas (1888) pp. 197–200.

P. orientalis (Pallas, 1766). Tate (1945) pp. 2–3 [as *P. orientalis peninsulae*].
‘

PETAURIDAE:

Pseudocheirus peregrinus (Boddaert, 1785). Thomas (1888) pp. 172–177 [including *P. peregrinus*, *P. occidentalis*, *P. cooki*].

P. herbertensis (Collett, 1884). Thomas (1888) pp. 170–172.

P. archeri (Collett, 1884). Thomas (1888) pp. 177–178.

Hemibelideus lemuroides (Collett, 1884). Thomas (1888) p. 170.

Petropseudes dahli (Collett, 1895). Collett (1897) pp. 329–332 [as *Pseudochirus dahli*].

Petaurus breviceps Waterhouse, 1839. Thomas (1888) pp. 156–158.

P. norfolcensis (Kerr, 1792). Thomas (1888) pp. 153–155 [as *P. sciureus*]. Calaby (1966) p. 24 [differences between *P. norfolcensis* and *P. breviceps* discussed under *P. breviceps*].

P. australis Shaw, 1791. Thomas (1888) pp. 151–153.

Schoinobates volans (Kerr, 1792). Thomas (1888) pp. 163–166 [including both *Petauroides volans* var. *typicus*, and var. *minor*]. Calaby (1966) pp. 26–27 [colour variation].

Gymnobelideus leadbeateri McCoy, 1867. Brazenor (1932) pp. 106–109.

Dactylopsila trivirgata Gray, 1858. Thomas (1888) pp. 159–161.

BURRAMYIDAE:

Acrobates pygmaeus (Shaw, 1793). Thomas (1888) pp. 136–138.

Cercartetus concinnus (Gould, 1845). Wakefield (1963) pp. 99–116.

C. nanus (Desmarest, 1818). Wakefield (1963) pp. 99–116.

C. lepidus (Thomas, 1888). Wakefield (1963) pp. 99–116.

C. caudatus (Milne-Edwards). Mjöberg (1916) pp. 13–20 [as *Eudromicia macrura*]. Wakefield (1963) pp. 99–116.

Burramys parvus Broom, 1896. Ride (1956) pp. 413–429. Lane & Richards (1967) p. 30, pl. 3.

TARSIPEDIDAE:

Tarsipes spencerae Gray, 1842. Thomas (1888) pp. 132–135.

PHASCOLARCTIDAE:

Phascolarctos cinereus (Goldfuss, 1817). Thomas (1888) pp. 209–212.

VOMBATIDAE:

Vombatus ursinus (Shaw, 1800). Thomas (1888) pp. 213–217 [including both *Phascolomys ursinus* and *P. mitchelli*]. Jones (1923–5) pp. 264–266 [as *P. mitchelli*]. Merrilees (1968) pp. 401–414 [as *V. hirsutus*].

Lasiorhinus latifrons (Owen, 1845). Thomas (1888) pp. 217–219 [as *P. latifrons*]. Jones (1923–5) pp. 266–270. Crowcroft (1968) pp. 383–397. Merrilees (1968) pp. 401–414.

L. gillespiei (De Vis, 1900). De Vis (1900) pp. 14–16 [as *Phascolomys gillespiei*]. Crowcroft (1968) pp. 392–397.

L. barnardi Longman, 1939. Longman (1939) pp. 283–287 [as *L. latifrons barnardi*]. Crowcroft (1968) pp. 394–396.

PERAMELIDAE:

Isoodon obesulus (Shaw, 1797). Jones (1923–5) pp. 138–145 [combine descriptions of *I. obesulus* and *I. nauticus*].

I. macrourus (Gould, 1842). Thomas (1888) pp. 234–235. Johnson (1964) pp. 446–450.

I. auratus (Ramsay, 1887). Ramsay (1887a) pp. 551–553.

Perameles nasuta Geoffroy, 1804. Freedman (1967) pp. 147–165. Freedman & Joffe (1967) pp. 183–195. Lyne (1964) pp. 322–339.

P. gunnii Gray, 1838. Freedman (1967) pp. 147–165. Freedman & Joffe (1967a) pp. 197–212. Lyne (1951) pp. 587–598.

P. bougainville Quoy & Gaimard, 1824. Freedman (1967) pp. 147–165. Freedman & Joffe (1967a) pp. 197–212.

P. eremiana Spencer, 1897. Spencer (1897) pp. 9–10. Lyne (1952) pp. 636–640.

Echymipera rufescens (Peters & Doria, 1875). Tate (1948a) p. 334 [as *Echymipera rufescens australis*].

Chaeropus ecaudatus (Ogilby, 1838). Thomas (1888) pp. 250–252. Lyne (1952) pp. 643–647.

Macrotis lagotis (Reid, 1837). Jones (1923–5) pp. 154–157 [as *Thalacomys lagotis*]. Finlayson (1935) pp. 233–236 [as *Thalacomys lagotis sagitta*]. Troughton (1932a) pp. 227–231.

M. leucura (Thomas, 1887). Finlayson (1935) pp. 227–233 [as *Thalacomys minor* var. *miselius*]. Troughton (1932a) pp. 231–236 [as *T. minor* and *T. leucura*].

DASYURIDAE:

Dasyurus maculatus (Kerr, 1792). Thomas (1888) pp. 263–265.

D. viverrinus (Shaw, 1800). Green (1967) pp. 1–13. Thomas (1888) pp. 265–268.

D. geoffroii Gould, 1841. Thomas (1888) pp. 268–269.

D. hallucatus Gould, 1842. Thomas (1888) pp. 269–271.
 Johnson (1964) pp. 443–446 [including both *Satanellus hallucatus hallucatus* and *S. h. nesaeus*].

Sarcophilus harrisii (Boitard, 1841). Green (1967) pp. 1–13.

Phascogale tapoatafa (Meyer, 1793). Thomas (1888) pp. 294–296 [as *P. penicillata*].

P. calura Gould, 1844. Thomas (1888) pp. 296–297.

Dasyuroides byrnei Spencer, 1896. Spencer (1896) pp. 36–40. Mack (1961) pp. 214–216.

Dasycercus cristicauda (Krefft, 1867). Jones (1949) pp. 409–501.

Antechinus flavipes (Waterhouse, 1838). Wakefield & Warneke (1967) pp. 69–95.

A. stuartii Macleay, 1841. Wakefield & Warneke (1967) pp. 69–95.

A. bellus (Thomas, 1904). Johnson (1964) pp. 437–439.

A. godmani (Thomas, 1923). Wakefield & Warneke (1967) pp. 95–98.

A. apicalis (Gray, 1842). Thomas (1888) pp. 277–278. Morcombe (1967) pp. 102, 109–110

A. rosamondae Ride, 1964. Ride (1964) pp. 58–65.

A. swainsonii (Waterhouse, 1840). Wakefield & Warneke (1963) pp. 194–219.

A. minimus (Geoffrey, 1803). Wakefield & Warneke (1963) pp. 194–219.

A. macdonnellensis (Spencer, 1896). Spencer (1896) pp. 27–30. Ride (1964) pp. 62–64 [dental variation and bulla only].

A. maculatus Gould, 1851. Johnson (1964) pp. 440–442 [as *A. m. sinualis*].

Planigale ingrami (Thomas, 1906). Troughton (1928) pp. 282–285, 288 [as *P. ingrami brunneus*].

P. subtilissima (Lonnberg, 1913). Lonnberg (1913) pp. 9–10 [as *Phascogale subtilissima*—description modified by the note on the age of the specimen by Tate (1947) p. 134].

P. tenuirostris Troughton, 1928. Troughton (1928) pp. 285–288.

Sminthopsis murina (Waterhouse, 1838). Jones (1923–5) pp. 117–118.

S. leucopus (Gray, 1842). Thomas (1888) pp. 302–303.

S. rufigenis Thomas, 1922. Thomas (1888) pp. 300–302 [as *S. virginiae*].

S. nitela Collett, 1897. Collett (1897) pp. 334–335.

S. longicaudata Spencer, 1909. Spencer (1909) pp. 449–451.

S. psammophila Spencer, 1895. Spencer (1896) pp. 35–36.

S. crassicaudata (Gould, 1844). Jones (1923–5) pp. 111–115. Finlayson (1933a) pp. 197–199 [including both *S. c. centralis* and *S. c. ferruginea*].

S. macroura (Gould, 1845). Troughton (1964) pp. 310–311.

S. froggatti (Ramsay, 1887). Finlayson (1933a) pp. 199–200 [as *S. larapinta*].

S. granulipes Troughton, 1932. Troughton (1932) pp. 350–352.

S. hirtipes Thomas, 1898. Troughton (1964) p. 314.

Antechinomys spenceri Thomas, 1906. Jones (1923–5) pp. 199–122.

A. laniger (Gould, 1856). Thomas (1888) pp. 309–310.

Myrmecobius fasciatus Waterhouse, 1836. Thomas (1888) pp. 311–315.

THYLACINIDAE:

Thylacinus cynocephalus (Harris, 1808). Thomas (1888) pp. 255–257.

NOTORYCTIDAE:

Notoryctes typhlops (Stirling, 1889). Jones (1923–5) pp. 128–131. Stirling (1891) pp. 158–184.

RODENTIA

MURIDAE:

Rattus fuscipes (Waterhouse, 1839). Troughton (1920) pp. 119–122 [as *Epimys assimilis*; note that the "*E. fuscipes*" of the comparison is actually *R. lutreolus*]. Finlayson (1960a) pp. 123–147 [as *R. greyi*]. Taylor (1961) pp. 9–18, 25–28 [as *R. assimilis*— growth stages and sexual characters]. Taylor & Horner (1967) [description of a neotype on pp. 8–9]. Brazenor (1936a) pp. 66–69 [including both *R. assimilis* and *R. greyi ravus*].

R. lutreolus (Gray, 1841). Waite (1900) pp. 190–193 [as *Mus fuscipes*]. Green (1967a) pp. 4–19 [under *R. l. velutinus*]. Brazenor (1936a) pp. 69–70.

R. sordidus (Gould, 1858). Thomas (1904) p. 599 [as *Mus colletti*]. McDougall (1944) pp. 59–60 [as *R. conatus*].

R. villosissimus (Waite, 1897). Finlayson (1939) pp. 88–94.

R. leucopus (Gray, 1867). Tate (1951) pp. 334–335 [as *R. l. leucopus*].

R. tunneyi (Thomas, 1904). Finlayson (1942) pp. 246–247 [under *R. culmorum*]. Johnson (1964) pp. 488, 490.

Hydromys chrysogaster Geoffroy, 1804. Tate (1951) pp. 228–234 [including *H. c. chrysogaster*, *H. c. fulvolavatus*, *H. c. fuliginosus* and *H. c. reginae*]. Brazenor (1936a) pp. 63–66.

Xeromys myoides Thomas, 1889. Tate (1951) pp. 225–226.

Mesembriomys macrurus (Peters, 1876). Peters (1876) pp. 355–357 [as *Hapalotis macrura*]. Ramsay (1887) pp. 1153–1154 [as *Hapalotis boweri*]. Collett (1897) pp. 321–322 [as *Conilurus boweri*].

M. gouldii (Gray, 1843). Finlayson (1961) pp. 149–162 [*as M. hirsutus*].

Conilurus albipes (Lichtenstein, 1829). Tate (1951) pp. 270–271.

C. penicillatus (Gould, 1842). Tate (1951) pp. 271–272 [including *C. p. penicillatus*, *C. p. hemileucurus*, and *C. melibus*].

Leporillus conditor (Sturt, 1848). Troughton (1923) pp. 24–27. Finlayson (1939) pp. 111–114.

L. apicalis (Gould, 1853). Finlayson (1941) pp. 228–232.

Notomys mitchellii (Ogilby, 1838). Finlayson (1939a) pp. 358–364 [as *N. mitchelli macropus*].

N. alexis Thomas, 1922. Finlayson (1940) pp. 125–134.

N. fuscus (Jones, 1925). Finlayson (1939) pp. 108–110 [as *N. cervinus* of Waite]. Finlayson (1960) pp. 80–82.

N. cervinus (Gould, 1853). Finlayson (1939) pp. 103–108 [as *N. aistoni*]. Finlayson (1960) p. 80.

N. megalotis Iredale & Troughton, 1934. Tate (1951) p. 264.

N. longicaudatus (Gould, 1844). Brazenor (1934) p. 84. Tate (1951) pp. 264–265 [for measurements see Finlayson (1940) p. 135].

N. amplus Brazenor, 1936. Brazenor (1936) pp. 7–8.

N. aquilo Thomas, 1921. Johnson (1959) pp. 186–187. Johnson (1964) pp. 497–500 [both references under *Notomys carpentarius*].

N. mordax Thomas, 1922. Thomas (1922) p. 317.

Zyzomys argurus (Thomas, 1889). Johnson (1964) pp. 484–485, 488 [under *Z. a. indutus*].

Z. woodwardi (Thomas, 1909). Thomas (1909) pp. 373–374 [as *Laomys woodwardi*]. Johnson (1964) pp. 484–486, 488.

Z. pedunculatus (Waite, 1896). Finlayson (1941) pp. 223–224 [as *Laomys pedunculatus*].

Mastacomys fuscus Thomas, 1882. Finlayson (1933) pp. 125–129. Green (1968) pp. 12–18.

Pseudomys forresti (Thomas, 1906). Finlayson (1941) pp. 220–222 [as *Pseudomys (Leggadina) waitei*].

P. delicatulus (Gould, 1842). Tate (1951) pp. 251–252 [as *Leggadina delicatula* and *L. patria*]. Johnson (1964) pp. 493, 495–496 [as *Leggadina d. delicatula* and *L. d. mimula*].

P. hermannsburgensis (Waite, 1896). Finlayson (1941) pp. 215–220.

P. novaehollandiae (Waterhouse, 1843). Keith & Calaby (1968) pp. 45–48.

P. albocinereus (Gould, 1845). Finlayson (1944) pp. 210–224 [under *P. (Gyomys) apodemoides*].

P. fumeus Brazenor, 1934. Brazenor (1934a) pp. 158–159.

P. occidentalis Tate, 1951. Tate (1951) p. 246.

P. praeconis Thomas, 1910. Tate (1951) p. 248. Ride & Tyndale-Biscoe (1962) pp. 77–78 [as *Thetomys praeconis*].

P. gouldii (Waterhouse, 1839). Brazenor (1936a) pp. 71–72. Finlayson (1939a) pp. 354–357 [as *P. (P.) rawlinnae*].

P. fieldi (Waite, 1896). Waite (1896) pp. 403–404 [as *Mus fieldi*].

P. australis Gray, 1832. Finlayson (1939) pp. 94–101 [as *P. (P.) minnie*].

P. higginsi (Trouessart, 1899). Green (1968) pp. 1–11.

P. oralis Thomas, 1921. Thomas (1921) pp. 621–622 [as *P. australis oralis*].

P. shortridgei (Thomas, 1907). Thomas (1907) pp. 765–766 [as *Mus shortridgei*].

P. desertor Troughton, 1932. Finlayson (1941) pp. 224–228 [as *P. (Thetomys) nanus*].

P. gracilicaudatus (Gould 1845). Tate (1951) p. 249 [including both *P. (T.) gracilicaudatus* and *P. (T.) gracilicaudatus ultra*]. Troughton (1939) p. 281 [as *Thetomys gracilicaudatus ultra*].

P. nanus (Gould, 1858). Tate (1951) pp. 247–248 [including both *P. (T.) nanus* and *P. (T.) ferculinus*].

Uromys caudimaculatus (Krefft, 1867). Troughton (1929) pp. 98–99 [as *U. macropus exilis*]. Tate (1951) pp. 309–310.

Melomys cervinipes (Gould, 1852). McDougall (1944) pp. 64–65. Tate (1951) pp. 292–294 [as *M. cervinipes cervinipes*, *M. c. bunya*, and *M. c. eboreus*].

M. lutillus (Thomas, 1913). Thomas (1913) p. 216 [as *Uromys lutillus*]. Thomas (1924) p. 298 [as *M. australius*].

M. littoralis (Lonnberg, 1916). Lonnberg (1916) pp. 5–6 [as *Uromys littoralis*]. McDougall (1944) pp. 63–64.

CHIROPTERA

MEGADERMATIDAE:

Macroderma gigas (Dobson, 1880). Jones (1923–5) pp. 440–444. Douglas (1962) pp. 59–61 [the dark *saturata* form].

VESPERTILIONIDAE:

Nyctophilus timoriensis (Geoffrey, 1806). Tate (1941) pp. 592–593 [as "*N. timoriensis* group"].

N. geoffroyi Leach, 1821. Jones (1923–5) pp. 436–440.

N. bifax Thomas, 1915. Tate (1941) p. 593 [as "*N. bifax* group"].

N. arnhemensis Johnson, 1959. Johnson (1959) pp. 184–185. Johnson (1964) pp. 478–481.

N. walkeri Thomas, 1892. Thomas (1892) p. 406. Tate (1941) p. 594 [measurements of skull in last paragraph of "*N. microtis* group"].

Miniopterus schreibersii (Kuhl, 1819). Jones (1923–5) pp. 430–433.

M. australis Tomes, 1858. Tate (1941) pp. 568, 572.

Eptesicus pumilus (Gray, 1841). Jones (1923–5) pp. 420–423. Green (1965) pp. 1–16.

Chalinolobus gouldii (Gray, 1841). Jones (1923–5) pp. 416–419.

C. morio (Gray, 1841). Jones (1923–5) pp. 413–416. Ryan (1966) pp. 87–88.

C. picatus (Gould, 1852). Ryan (1966) pp. 87–88.

C. dwyeri Ryan, 1966. Ryan (1966) pp. 88–89.

C. rogersi Thomas, 1909. Johnson (1964) pp. 476–477 [as *C. picatus nigrogriseus*].

Pipistrellus tasmaniensis (Gould, 1858). Tate (1942) pp. 250–251 [including *P. tasmaniensis* and *P. t. krefftii*].

P. javanicus (Gray, 1838). Dobson (1878) pp. 226–228 [as *Vesperugo abramus*].

P. tenuis (Temminck, 1840). Dobson (1878) p. 226 [as *Vesperugo tenuis*].

P. papuanus (Peters & Doria, 1881). Peters & Doria (1881) pp. 696–697 [as *Vesperugo papuanus*].

Myotis adversus (Horsfield, 1824). Jones (1923–5) pp. 408–411.

M. australis (Dobson, 1878). Dobson (1878) pp. 317–318 [as *Vespertilio australis*].

Nycticeius rueppellii Peters, 1866. Dobson (1878) p. 263 [as *Scotophilus rueppellii*].

N. greyi (Gould, 1858). Jones (1923–5) pp. 424–429 [including both *Scoteinus balstoni* and *S. greyi*].

N. influatus (Thomas, 1924). Thomas (1924a) p. 540.

Phoniscus papuensis (Dobson, 1878). Hill (1965) p. 551.

RHINOLOPHIDAE:

Rhinolophus megaphyllus Gray, 1834. Andersen (1905) pp. 79–80 [as *R. megaphyllus* form *typicus*].

R. philippinensis Waterhouse, 1843. Tate (1952a) pp. 1–3 [as *R. maros robertsi*].

HIPPOSIDERIDAE:

Hipposideros ater Templeton, 1848. Hill (1963) pp. 30–33. Johnson (1959) pp. 183–184 [as *H. bicolor gilberti*]. Johnson (1964) pp. 471–473 [as *H. bicolor gilberti*].

H. galeritus Cantor, 1846. Hill (1963) pp. 52–58.

H. diadema (Geoffroy, 1813). Hill (1963) pp. 108–111.

H. semoni Matschie, 1903. Hill (1963) pp. 84–86.

H. stenotis Thomas, 1913. Hill (1963) pp. 86–87.

Rhinonicteris aurantius (Gray 1845). Miller (1907) pp. 114–115. Jones (1923–5) pp. 446–449.

MOLOSSIDAE:

Tadarida australis (Gray, 1839). Jones (1923–5) pp. 387–393 [under *Nyctinomus australis* and *N. a. atratus*]. Hill (1961) pp. 35, 37–39.

T. jobensis (Miller, 1902). Jones (1923–5) pp. 397–399 [under *Chaerophon plicatus*]. Hill (1961) pp. 54–55.

T. loriae (Thomas, 1897). Felten (1964) pp. 6–8 [under *T. loriae ridei*, *T. loriae loriae*, and *T. loriae cobourgiana*].

T. planiceps (Peters, 1866). Felten (1964) pp. 3–5. Jones (1923–5) pp. 393–396 [under *Nyctinomus petersi*].

T. norfolkensis (Gray, 1839). Felten (1964) p. 6.

EMBALLONURIDAE:

Taphozous georgianus Thomas, 1915. Troughton (1925) pp. 336–341 [but gular pouch sometimes present].

T. flaviventris Peters, 1867. Troughton (1925) pp. 315–321, 341 [as *Saccolaimus flaviventris*].
T. australis Gould, 1854. Troughton (1925) pp. 332–336, 340–341.
T. nudicluniatus De Vis, 1905. Troughton (1925) pp. 325–328, 340–341 [as *Saccolaimus nudicluniatus*].

T. mixtus (Troughton, 1925). Troughton (1925) pp. 322–325, 340–341 [as *Saccolaimus mixtus*]. Tate (1952) p. 606.

PTEROPODIDAE:

Pteropus scapulatus Peters, 1862. Andersen (1912) pp. 403–407. Johnson (1964) pp. 468–470.

P. poliocephalus Temminck, 1825. Andersen (1912) pp. 397–402.

P. alecto Temminck, 1837. Andersen (1912) pp. 370–374 [as *Pteropus gouldi*]. Johnson (1964) pp. 467–469. [as *P. alecto gouldii*].

P. conspicillatus Gould, 1850. Andersen (1912) pp. 378–381.
Dobsonia moluccense (Quoy & Gaimard, 1830). Andersen (1912) pp. 464–467 [including *D. moluccensis* and *D. magna*].

Nyctimene robinsoni Thomas, 1904. Andersen (1912) pp. 714, 718, 720, 722.

N. albiventer (Gray, 1863). Andersen (1912) pp. 698–701 [including *N. albiventer* and *N. papuanus*].

Syconycteris australis (Peters, 1867). Andersen (1912) pp. 781–784.

Macroglossus lagochilus Matschie, 1899. Andersen (1912) pp. 762–770.

CARNIVORA

CANIDAE:

Canis familiaris Linnaeus, 1758. Jones (1923–5) pp. 349–356 [as *C. f. dingo*].

OTARIIDAE:

Neophoca cinerea (Péron & Lesueur, 1816). Jones (1923–5) pp. 362–373.
Arctocephalus doriferus Jones, 1925. Jones (1923–5) pp. 373–377.
A. forsteri (Lesson, 1828). Jones (1923–5) pp. 377–379.

PHOCIDAE:

Mirounga leonina (Linnaeus, 1758). Lydekker (1909) pp. 600–606. King (1964) pp. 78–79.

MONOTREMATA

TACHYGLOSSIDAE:

Tachyglossus aculeatus (Shaw, 1792). Thomas (1888) pp. 377–382, 384 [as *Echidna aculeata* including vars *lawesi, typica* and *setosa*].

ORNITHORHYNCHIDAE:

Ornithorhynchus anatinus (Shaw, 1799). Thomas (1888) pp. 385–391.

REFERENCES

The references listed here are confined to those containing the descriptions recommended above; they do not include the works in which the names of taxa were first published unless no better description is available. Such original works are cited in the list above through including authorship and date as a part of each name and full references to them may be found in

IREDALE, T., and TROUGHTON, E. LE G. (1934): A check-list of the mammals recorded from Australia. *Mem. Aust. Mus.* 6: 1–122.

RIDE, W. D. L. (1964): A list of mammals described from Australia between the years 1933 and 1963 (comprising newly proposed names and additions to the Australian faunal list). *Aust. Mammal Soc. Bull.* 7 (Supplement): 1–15.

ANDERSEN, K. (1905): On some bats of the genus *Rhinolophus*, with remarks on their mutual affinities, and descriptions of twenty-six new forms. *Proc. zool. Soc. Lond.* 1905 2: 75–145.

ANDERSEN, K. (1912): Catalogue of the Chiroptera in the collection of the British Museum (2 Ed.). Vol. 1: Megachiroptera. 854 pp. London: British Museum (Natural History).

BRAZENOR, C. W. (1932): A re-examination of *Gymnobelideus leadbeateri* McCoy. *Aust. Zool.* 7: 106–109.

BRAZENOR, C. W. (1934): A revision of the Australian jerboa mice. *Mem. natn. Mus., Melb.* no. 8: 74–89.

BRAZENOR, C. W. (1934a): A new species of mouse, *Pseudomys* (*Gyomys*), and a record of the broad-toôthed rat, *Mastacomys*, from Victoria. *Mem. natn. Mus., Melb.* no. 8: 158–161.

BRAZENOR, C. W. (1936): Two new rats from Central Australia. *Mem. natn. Mus., Melb.* no. 9: 5–8.

BRAZENOR, C. W. (1936a): Muridae recorded from Victoria. *Mem. natn. Mus., Melb.* no. 10: 62–85.

CALABY, J. H. (1957): A new record of the scaly-tailed possum (*Wyulda squamicaudata* Alexander). *W. Aust. Nat.* 5: 186–191.

CALABY, J. H. (1966): Mammals of the Upper Richmond and Clarence Rivers, New South Wales. *Tech. Pap. Div. Wildl. Surv. C.S.I.R.O. Aust.* no. 10: 1–55.

CARLSSON, A. (1915): Zur morphologie des *Hypsiprymnodon moschatus*. *K. svenska VetenskAkad. Handl.* 52 no. 6: 1–48.

COLLETT, R. (1897): On a collection of mammals from North and North-west Australia. *Proc. zool. Soc. Lond.* 1897: 317–336.

CROWCROFT, P. (1968): Studies on the Hairy-nosed Wombat *Lasiorhinus latifrons* (Owen 1845). I. Measurements and Taxonomy. *Rec. S. Aust. Mus.* 15: 383–398.

DE VIS, C. W. (1900): A new species of hairy-nosed wombat. *Ann. Qd Mus.* no. 5: 14–16.

DOBSON, G. E. (1878): Catalogue of the Chiroptera in the collection of the British Museum. 567 pp. London: British Museum. (Reprinted 1966: Wheldon & Wesley.)

DOUGLAS, A. M. (1962): *Macroderma gigas saturata* (Chiroptera, Megadermatidae), a new subspecies from the Kimberley Division of Western Australia. *W. Aust. Nat.* 8: 59–61.

FELTEN, H. (1964): Zur Taxionomie indo-australischer Fledermause der Gattung *Tadarida* (Mammalia, Chiroptera). *Senckenberg. biol.* 45: 1–13.

FINLAYSON, H. H. (1927): Observations on the South Australian species of the subgenus, "Wallabia". *Trans. R. Soc. S. Aust.* 51: 363–377.

FINLAYSON, H. H. (1931): On mammals from the Dawson Valley, Queensland. Part I. *Trans. R. Soc. S. Aust.* **55**: 67-89.

FINLAYSON, H. H. (1932): *Caloprymnus campestris*. Its recurrence and characters. *Trans. R. Soc. S. Aust.* **56**: 148-167.

FINLAYSON, H. H. (1933): On *Mastacomys fuscus* Thomas. *Trans. R. Soc. S. Aust.* **57**: 125-129.

FINLAYSON, H. H. (1933a): On mammals from the Lake Eyre Basin. Part I. The Dasyuridae. *Trans. R. Soc. S. Aust.* **57**: 195-202.

FINLAYSON, H. H. (1935): On mammals from the Lake Eyre Basin. Part II. The Peramelidae. *Trans. R. Soc. S. Aust.* **59**: 227-236.

FINLAYSON, H. H. (1938): On a new species of *Potoroüs* (Marsupialia) from a cave deposit on Kangaroo Island, South Australia. *Trans. R. Soc. S. Aust.* **62**: 132-140.

FINLAYSON, H. H. (1939): On mammals from the Lake Eyre Basin. Pt. IV. The Monodelphia. *Trans. R. Soc. S. Aust.* **63**: 88-118.

FINLAYSON, H. H. (1939a): Records and descriptions of Muridae from Ooldea, South Australia. *Trans. R. Soc. S. Aust.* **63**: 354-364.

FINLAYSON, H. H. (1940): On Central Australian mammals: Part 1. The Muridae. *Trans. R. Soc. S. Aust.* **64**: 125-136.

FINLAYSON, H. H. (1941): On Central Australian mammals: Part II. The Muridae. *Trans. R. Soc. S. Aust.* **65**: 215-232.

FINLAYSON, H. H. (1942): A new *Melomys* from Queensland with notice of two other Queensland rats. *Trans. R. Soc. S. Aust.* **66**: 243-247.

FINLAYSON, H. H. (1943): A second specimen of *Wyulda squamicaudata* Alexander. *Trans. R. Soc. S. Aust.* **66**: 255-261.

FINLAYSON, H. H. (1943a): A new species of *Lagorchestes* (Marsupialia). *Trans. R. Soc. S. Aust.* **67**: 319-321.

FINLAYSON, H. H. (1944): A further account of the murid, *Pseudomys* (*Gyomys*) *apodemoides* Finlayson. *Trans. R. Soc. S. Aust.* **68**: 210-224.

FINLAYSON, H. H. (1958): On Central Australian mammals (with notice of related species from adjacent tracts). Part III — The Potoroinae. *Rec. S. Aust. Mus.* **13**: 235-302.

FINLAYSON, H. H. (1959): On *Bettongia cuniculus* Ogilby, 1838 (Marsupialia). *Trans. R. Soc. S. Aust.* **82**: 283-289.

FINLAYSON, H. H. (1960): Nomenclature of *Notomys* (Muridae) in the Lake Eyre Basin. *Trans. R. Soc. S. Aust.* **83**: 79-82.

FINLAYSON, H. H. (1960a): On *Rattus greyi* Gray and its derivatives. *Trans. R. Soc. S. Aust.* **83**: 123-147.

FINLAYSON, H. H. (1961): A re-examination of *Mesembriomys hirsutus* Gould 1842 (Muridae). *Trans. R. Soc. S. Aust.* **84**: 149-162.

FREEDMAN, L. (1967): Skull and tooth variation in the genus *Perameles*. Part 1: Anatomical features. *Rec. Aust. Mus.* **27**: 147-166.

FREEDMAN, L., and JOFFE, A. D. (1967): Skull and tooth variation in the genus *Perameles*. Part 2: Metrical features of *P. nasuta*. *Rec. Aust. Mus.* **27**: 183-195.

FREEDMAN, L., and JOFFE, A. D. (1967a): Skull and tooth variation in the genus *Perameles*. Part 3: Metrical features of *P. gunnii* and *P. bougainville*. *Rec. Aust. Mus.* **27**: 197-212.

GREEN, R. H. (1965): Observations on the Little Brown Bat *Eptesicus pumilus* Gray in Tasmania. *Rec. Queen Vict. Mus.* (n.s.) no. 20: 1-16.

GREEN, R. H. (1967): Notes on the Devil (*Sarcophilus harrisi*) and the Quoll (*Dasyurus viverrinus*) in north-eastern Tasmania. *Rec. Queen Vict. Mus.* no. **27**: 1-13.

GREEN, R. H. (1967a): The murids and small dasyurids in Tasmania. Parts 1 and 2. *Rec. Queen Vict. Mus.* no. 28: 1-19.

GREEN, R. H. (1968): The murids and small dasyurids in Tasmania. Parts 3 and 4. *Rec. Queen Vict. Mus.* no. 32: 1-19.

HILL, J. E. (1961): Indo-Australian bats of the genus *Tadarida*. *Mammalia* **25**: 29-56.

HILL, J. E. (1963): A revision of the genus *Hipposideros*. *Bull. Br. Mus. nat. Hist.* (*Zool.*) **11**: 1-129.

HILL, J. E. (1965): Asiatic bats of the genera *Kerivoula* and *Phoniscus* (Vespertilionidae), with a note on *Kerivoula aerosa* Tomes. *Mammalia* **29**: 524-556.

JOHNSON, D. H. (1959): Four new mammals from the Northern Territory of Australia. *Proc. biol. Soc. Wash.* **72**: 183–187.

JOHNSON, D. H. (1964): Mammals, in: *Records of the American–Australian Scientific Expedition to Arnhem Land,* **4**: 427–515. Ed. R. L. Specht, Melbourne: University Press.

JONES, F. WOOD (1923–25): *The Mammals of South Australia.* Adelaide, Govt. Print. Pt 1. Monotremes and the carnivorous marsupials: 1–131. Pt. 2 Bandicoots and the herbivorous marsupials: 132–270. Pt 3. Monodelphia: 271–458.

JONES, F. WOOD (1949): The study of a generalized marsupial (*Dasycercus cristicauda* Krefft). *Trans. zool. Soc. Lond.* **26**: 409–501.

KEITH, K., and CALABY, J. H. (1968): The New Holland Mouse, *Pseudomys novaehollandiae* (Waterhouse), in the Port Stephens District, New South Wales. *C.S.I.R.O. Wildl. Res.* **13**: 45–58.

KING, J. E. (1964): *Seals of the World.* London: British Museum (Natural History). 154 pp.

KIRSCH, J. A. W., and POOLE, W. E. (1967): Serological evidence for speciation in the Grey Kangaroo, *Macropus giganteus* Shaw 1790 (Marsupialia: Macropodidae). *Nature, Lond.* **215**: 1097–1098.

LANE, E. A., and RICHARDS, AOLA M. (1967): *Burramys parvus* Broom, a living fossil. *Helictite* **5**: 30–34.

LE SOUEF, A. S., and BURRELL, H. (1926): *The wild animals of Australasia.* With a chapter on bats by Ellis Le G. Troughton. London, George G. Harrap. pp. 388.

LONGMAN, H. A. (1939): A central Queensland Wombat. *Mem. Qd Mus.* **11**: 283–287.

LONNBERG, E. (1913): Results of Dr E. Mjöberg's Swedish Scientific Expeditions to Australia 1910–13. 1. Mammals. *K. svenska VetenskAkad. Handl.* **52** No. 1:1–10.

LONNBERG, E. (1916): Results of Dr E. Mjöberg's Swedish Scientific Expeditions to Australia 1910–13. 2. Mammals from Queensland. 1. List of mammals. *K. svenska VetenskAkad. Handl.* **52** No. 2: 3–11.

LYDEKKER, R. (1909): On the skull characters in the Southern Sea-Elephant. *Proc. zool. Soc. Lond.* **1909**: 600–606.

LYNE, A. G. (1951): Notes on external characters of the Barred Bandicoot (*Perameles gunnii* Gray), with special reference to the pouch young. *Proc. zool. Soc. Lond.* **121**: 587–598.

LYNE, A. G. (1952): Notes on external characters of the pouch young of four species of bandicoots. *Proc. zool. Soc. Lond.* **122**: 625–649.

LYNE, A. G. (1964): Observations on the breeding and growth of the marsupial *Perameles nasuta* Geoffroy, with notes on other bandicoots. *Aust. J. Zool.* **12**: 322–339.

LYNE, A. G., and VERHAGEN, A. M. W. (1957): Growth of the marsupial *Trichosurus vulpecula* and a comparison with some higher mammals. *Growth* **21**: 167–195.

MC DOUGALL, W. A. (1944): An investigation of the rat pest problem in Queensland canefields: 2 species and general habits. *Qd J. agric. Sci.* **1**(2): 48–78.

MACK, G. (1961): Mammals from south-western Queensland. *Mem. Qd Mus.* **13**: 213–229.

MERRILEES, D. (1968): Cranial and mandibular characters of modern mainland wombats (Marsupialia, Vombatidae) from a palaeontological viewpoint, and their bearing on the fossils called *Phascolomys parvus* by Owen (1872). *Rec. S. Aust. Mus.* **15**: 399–418.

MILLER, G. S. (1907): The families and genera of bats. *Bull. U.S. natn. Mus.* No. 57: 1–282.

MJÖBERG, E. (1916): Results of Dr E. Mjöberg's Swedish Scientific Expeditions to Australia 1910–1913. 2 Mammals from Queensland. 2. On a new genus and species of marsupials. *K. svenska VetenskAkad. Handl.* **52**: no. 2: 13–20.

MORCOMBE, M. K. (1967): The rediscovery after 83 years of the Dibbler *Antechinus apicalis* (Marsupialia, Dasyuridae). *W. Aust. Nat.* **10**: 102–111.

PETERS, W. (1876): Über die von S.M.S. Gazelle gesammelten Säugethiere aus den Abtheilungen der Nager, Hufthiere, Sirenen, Cetaceen und Beutelthiere. *Mber. preuss. Akad. Wiss.* **1876**: 355–366.

PETERS, W., and DORIA, G. (1881): Enumerazione dei Mammiferi raccolti da O. Beccari, L. M. D'Albertis ed A. A. Bruijn nella Nuova Guinea propriamente detta. *Ann. Mus. Stor. nat. Genova* **16**: 664–707.

RAMSAY, E. P. (1887): Description of a new species of *Hapalotis*, (*H. boweri*) from north-west Australia. *Proc. Linn. Soc. N.S.W.* (2) 1: 1153–4.

RAMSAY, E. P. (1887a): Description of two new species of marsupials (*Perameles* and *Antechinus*) and of a new species of *Mus* (*M. burtoni*), from the neighbourhood of Derby, N.W.A. *Proc. Linn. Soc. N.S.W.* (2) 2: 551–553.

RIDE, W. D. L. (1956): The affinities of *Burramys parvus* Broom a fossil phalangeroid marsupial. *Proc. zool. Soc. Lond.* 127: 413–429.

RIDE, W. D. L. (1957): *Protemnodon parma* (Waterhouse) and the classification of related wallabies (*Protemnodon, Thylogale* and *Setonyx*). *Proc. zool. Soc. Lond.* 128: 327–346.

RIDE, W. D. L. (1964): *Antechinus rosamondae*, a new species of dasyurid marsupial from the Pilbara District of Western Australia; with remarks on the classification of *Antechinus*. *W. Aust. Nat.* 9: 58–65.

RIDE, W. D. L., and TYNDALE-BISCOE, C. H. (1962): Mammals, pp. 54–97, in Ride *et al.* (1962). The results of an Expedition to Bernier and Dorre Islands, Shark Bay, Western Australia, in July 1959. *Fauna Bull. Fish. Dep. West. Aust.* no. 2: 1–131.

ROTHSCHILD, W. (1904): Preliminary diagnosis of a new genus and species of kangaroo. *Novit. zool.* 10: 414.

ROTHSCHILD, Lord, and DOLLMAN, G. (1936): The genus *Dendrolagus. Trans. zool. Soc. Lond.* 21: 477–548.

RYAN, R. M. (1966): A new and some imperfectly known Australian *Chalinolobus* and the taxonomic status of African *Glauconycteris. J. Mammal.* 47: 86–91.

SPENCER, W. B. (1896): Mammalia, pp. 1–52, in *Report on the work of the Horn Scientific Expedition to Central Australia* Pt 2. — Zoology. Ed. Spencer, W. B., Melb.: Melville, Mullen and Slade.

SPENCER, W. B. (1897): Description of two new species of marsupials from Central Australia. *Proc. R. Soc. Vict.* (n.s.) 9: 5–11.

SPENCER, W. B. (1909): Description of a new species of *Sminthopsis. Proc. R. Soc. Vict.* (n.s.) 21: 449–451.

STIRLING, E. C. (1891): Description of a new genus and species of Marsupialia, "*Notoryctes typhlops*". *Trans. R. Soc. S. Aust.* 14: 154–187.

TATE, G. H. H. (1941): Results of the Archbold Expeditions. No. 40. Notes on vespertilionid bats of the subfamilies Miniopterinae, Murininae, Kerivoulinae, and Nyctophilinae. *Bull. Am. Mus. nat. Hist.* 78: 567–597.

TATE, G. H. H. (1942): Results of the Archbold Expeditions. No. 47. Reviews of the vespertilionine bats, with special attention to genera and species of the Archbold Collections. *Bull. Am. Mus. nat. Hist.* 80: 221–297.

TATE, G. H. H. (1945): Results of the Archbold Expeditions. No. 52. The marsupial genus *Phalanger. Am. Mus. Novit.* no. 1283: 1–41.

TATE, G. H. H. (1947): Results of the Archbold Expeditions. No. 56. On the anatomy and classification of the Dasyuridae (Marsupialia). *Bull. Am. Mus. nat. Hist.* 88: 101–155.

TATE, G. H. H. (1948): Results of the Archbold Expeditions. No. 59. Studies on the anatomy and phylogeny of the Macropodidae (Marsupialia). *Bull. Am. Mus. nat. Hist.* 91: 233–351.

TATE, G. H. H. (1948a): Results of the Archbold Expeditions. No. 60. Studies in the Peramelidae (Marsupialia). *Bull. Am. Mus. nat. Hist.* 92: 313–346.

TATE, G. H. H. (1951): Results of the Archbold Expeditions. No. 65. The rodents of Australia and New Guinea. *Bull. Am. Mus. nat. Hist.* 97: 183–430.

TATE, G. H. H. (1952): Results of the Archbold Expeditions. No. 66. Mammals of Cape York peninsula, with notes on the occurrence of rain forest in Queensland. *Bull. Am. Mus. nat. Hist.* 98: 563–616.

TATE, G. H. H. (1952a): Results of the Archbold Expeditions. No. 67. A new Thinolophus from Queensland (Mammalia, Chiroptera). *Am. Mus. Novit.* no. 1578: 1–3.

TAYLOR, J. M. (1961): Reproductive biology of the Australian Bush Rat *Rattus assimilis. Univ. Calif. Publs Zool.* 60: 1–66.

TAYLOR, J. M., and HORNER, B. E. (1967): Results of the Archbold Expeditions. No. 88. The historical misapplication of the name *Mus fuscipes* and a systematic re-evaluation of *Rattus lacus* (Rodentia, Muridae). *Am. Mus. Novit.* no. 2281: 1–14.

THOMAS, O. (1887): On the wallaby commonly known as *Lagorchestes fasciatus. Proc. zool. Soc. Lond.* **1886**: 544–547.

THOMAS, O. (1888): *Catalogue of the Marsupialia and Monotremata in the collection of the British Museum (Natural History).* London, British Museum (Natural History). xiii, 401 pp.

THOMAS, O. (1892): Description of a third species of the genus *Nyctophilus. Ann. Mag. nat. Hist.* (6) **9**: 405–6.

THOMAS, O. (1904): New species of *Pteropus, Mus*, and *Pogonomys* from the Australian Region. *Novit. zool.* **11**: 597–600.

THOMAS, O. (1907): List of further collections of mammals from Western Australia, including a series from Bernier Island, obtained for Mr. W. E. Balston; with field-notes by the collector, Mr. G. C. Shortridge. *Proc. zool. Soc. Lond.* **1906**: 763–777.

THOMAS, O. (1909): On the N. Australian rats referred to the genus *Mesembriomys. Ann. Mag. nat. Hist.* (8) **3**: 272–374.

THOMAS, O. (1913): Some new species of *Uromys. Ann. Mag. nat. Hist.* (8) **12**: 212–217.

THOMAS, O. (1921): On three new Australian rats. *Ann. Mag. nat. Hist.* (9) **8**: 618–622.

THOMAS, O. (1922): Two new Jerboa-rats (*Notomys*). *Ann. Mag. nat. Hist.* (9) **9**: 315–317.

THOMAS, O. (1923): [Exhibition of a new rock-kangaroo: description of *Petrogale godmani.*] *Proc. zool. soc. Lond.* **1923**: 177–178.

THOMAS, O. (1924): Some new Australian Muirdae. *Ann. Mag. nat. Hist.* (9) **13**: 296-299.

THOMAS, O. (1924a): A new *Scoteinus* from Queensland. *Ann. Mag. nat. Hist.* (9) **13**: 540.

TROUGHTON, E. LE G. (1920): Notes on Australian mammals, No. 1. *Rec. Aust. Mus.* **13**: 118–122.

TROUGHTON, E. LE G. (1923): A revision of the rats of the genus *Leporillus* and the status of *Hapalotis personata* Krefft. *Rec. Aust. Mus.* **14**: 23–41.

TROUGHTON, E. LE G. (1925): A revision of the genera *Taphozous* and *Saccolaimus* (Chiroptera) in Australia and New Guinea, including a new species, and a note on two Malayan forms. *Rec. Aust. Mus.* **14**: 313–341.

TROUGHTON, E. LE G. (1928): A new genus, species, and subspecies of marsupial mice (family Dasyuridae). *Rec. Aust. Mus.* **16**: 281–288.

TROUGHTON, E. LE G. (1929): New forms of mosaic-tailed rats (*Melomys* and *Uromys*) from Hinchinbrook Island, Queensland. *Aust. Zool.* **6**: 96–99.

TROUGHTON, E. LE G. (1932): A new species of fat-tailed Marsupial mouse, and the status of *Antechinus froggatti* Ramsay. *Rec. Aust. Mus.* **18**: 349–353.

TROUGHTON, E. LE G. (1932a): A revision of the rabbit-bandicoots. Family Peramelidae, genus *Macrotis. Aust. Zool.* **7**: 219–236.

TROUGHTON, E. LE G. (1939): Queensland rats of economic importance and new forms of *Rattus* and *Thetomys. Rec. Aust. Mus.* **20**: 278–281.

TROUGHTON, E. LE G. (1964): A review of the marsupial genus *Sminthopsis* (Phascogalinae) and diagnosis of new forms. *Proc. Linn. Soc. N.S.W.* **89**: 307–321.

WAITE, E. (1896): Muridae, pp. 393–409. In *Report on the work of the Horn Scientific Expedition to Central Australia*, Pt 2. Zoology. Ed. Baldwin Spencer. Melbourne: Melville, Muller, and Slade.

WAITE, E. R. (1900): An extended description of *Mus fuscipes*, Waterhouse. *Rec. Aust. Mus.* **3**: 190–193.

WAITE, E. R. (1901): A description of *Macropus isabellinus*, Gould. *Rec. Aust. Mus.* **4**: 131–134

WAKEFIELD, N. A. (1963): The Australian pigmy-possums. *Vict. Nat., Melb.* **80**: 99–116.

WAKEFIELD, N. A. (1967): Some taxonomic revision in the Australian marsupial genus *Bettongia* (Macropodidae), with description of a new species. *Vict. Nat., Melb.* **84**: 8–22.

WAKEFIELD, N. A., and WARNEKE, R. M. (1963): Some revision in *Antechinus* (Marsupialia)—1. *Vict. Nat., Melb.* **80**: 194–219.

WAKEFIELD, N. A., and WARNEKE, R. M. (1967): Some revision in *Antechinus* (Marsupialia)—2. *Vict. Nat., Melb.* **84**: 69–99.

Appendix III Information on the drawings of native mammals *

PLATE 1 *Lagostrophus fasciatus*: Banded Hare-wallabies in captivity at Mr N. A. Beeck's property at Katanning; the stock was originally captured on Bernier Island, Shark Bay, W.A., by Drs C. H. Tyndale-Biscoe and W. D. L. Ride and bred by Mr N. A. Beeck.

PLATE 2 *Burramys parvus*: Two postures of the male *Burramys* captured at Mount Hotham, Vic. in 1966 by Dr K. Shortman and Mr D. Jamieson; based on photographs supplied by Fisheries and Wildlife Department, Victoria.

PLATE 3 *Wyulda squamicaudata*: In the foreground a female Scaly-tailed Possum with young; in the background the same juvenile which, at the time of illustration, had grown into a young adult female; animals captured by Mr W. H. Butler at Kalumburu, Kimberley, W.A.

PLATE 4 *Antechinus apicalis*: Postures of a male and female Dibbler; animals captured by Mr M. Morcombe near Mount Many Peaks, W.A.

PLATE 5 *Macropus parma*: Two postures of Parma Wallabies based on photographs taken in the wild by Dr J. Flux on Kawau Island, New Zealand; detail added from captive animals from Kawau held at the South Perth Zoological Gardens.

PLATE 6 *Macropus fuliginosus*: A boomer (reclining) and two does of the Western Grey Kangaroo; the male is in captivity at Yanchep Park, W.A., and the females in the yards of the Zoology Department, University of W.A.

PLATE 7 *Macropus irma*: Two adults and a juvenile Western Brush Wallaby; the studies of adults are of captive animals at the South Perth Zoological Gardens, the juvenile is in captivity at Yanchep Park, W.A.; the vegetation shown is sand-plain heath at Pingelly, W.A.

PLATE 8 *Macropus eugenii*: Tammars in captivity at the Zoology Department, University of W.A.; the vegetation is Tammar Scrub (*Casuarina*) at Tuttanning near Pingelly, W.A.

PLATE 9 *Setonix brachyurus*: Rottnest Quokkas in the yards of the Zoology Department, University of W.A.; the vegetation is typical of the coastal dunes.

PLATE 10 *Onychogalea unguifera*: Two studies of a male Karrabul, or Northern Nail-tailed Wallaby, in captivity at Mr N. A. Beeck's property at Katanning; the animal was captured as a juvenile at Kalumburu, in the north Kimberley, by Mr W. H. Butler. Postures are from photographs with detail and texture from life.

PLATE 11 *Lagorchestes conspicillatus*: Studies of Spectacled Hare-wallabies in the yards of the Zoology Department, University of W.A.; the animals were captured on Barrow Island, W.A., by Mr W. H. Butler; the background is based on photographs of Barrow Island by Dr B. R. Wilson; the action posture is from a photograph.

*When it is said here that an animal was held captive, but no specific place of captivity is mentioned, it is to be understood that the animal was kept and observed by Mrs Fry. Whenever photography is said to provide the basis of a drawing, the photographer's name is given unless it is W. D. L. Ride. Where there is no mention of photography, the drawings are from life (see p. x).

PLATE 12   *Petrogale rothschildi*: Studies of a female Rothschild's Rock-wallaby captured by Dr J. W. Shield near Woodstock Station in the Pilbara District, W.A.; the animal is in captivity in the Zoology Department, University of W.A.; the environment figured is from photographs and is that inhabited by the species on Rosemary Island in the Dampier Archipelago, W.A.

PLATE 13   *Dendrolagus goodfellowi*: A male of a New Guinea species of tree-kangaroo in captivity at the Baiyer River Sanctuary, near Mount Hagen, Highlands of New Guinea; from photographs.

PLATE 14   *Bettongia penicillata*: Woylies, or Brush-tailed Bettongs, in captivity at Mr N. A. Beeck's property at Katanning, W.A.; the stock came originally from near Pingelly, W.A. and were bred by Mr Beeck.

PLATE 15   *Trichosurus vulpecula*: A Brush Possum showing one of the common colour forms of south-western Australia; this variety is grey with a white tail-tip.

PLATE 16   *Phalanger maculatus*: A juvenile male Spotted Cuscus in captivity at Baiyer River Sanctuary, near Mount Hagen, Highlands of New Guinea; from photographs.

PLATE 17   *Pseudocheirus peregrinus*: Captive Ringtails collected by Mr W. H. Butler from near Busselton, South West of Western Australia; the south-western form of the Common Ringtail is very dark in colour.

PLATE 18   *Petaurus breviceps*: Several postures of a captive Sugar Glider from Kalumburu on the north coast of Kimberley, W.A., collected by Mr W. H. Butler; the flight posture is from a photograph taken of an enforced glide in captivity.

PLATE 19   *Gymnobelideus leadbeateri*: Several postures drawn from photographs supplied by the Fisheries and Wildlife Department, Victoria.

PLATE 20   *Dactylopsila trivirgata*: Two postures of a Striped Possum from New Guinea, drawn from photographs of an animal in captivity at the home of Mr Roy D. McKay, Port Moresby; detail added from a skin in the Western Australian Museum collected by Dr J. A. Thomson near Innisfail, northern Queensland.

PLATE 21   *Cercartetus concinnus*: Three postures of a Pigmy Possum from Busselton in the south-west of W.A.; the animal was collected by Mr R. A. Breeden; the postures are from photographs with detail added from the captive animal.

PLATE 22   *Tarsipes spencerae*: A female Honey Possum, and her two young; the three postures of the adult are all of the same female which was kept in captivity from the time that the young were small and hairless until they were almost as large as the parent; the young ones shown in the suckling posture are drawn as they appeared soon after they emerged from the pouch for the first time; the female was collected by Mr B. Stagg at Balcatta, a suburb of Perth.

PLATE 23   *Phascolarctos cinereus:* Koalas in captivity at Yanchep Park, W.A.; the stock is of mixed origins.

PLATE 24   *Vombatus ursinus*: Postures drawn from wombats captive at the South Perth Zoo; the animals came from Taronga Park Zoo and are probably from mainland Australia; the background is from photographs taken at Brindabella, A.C.T.

PLATE 25   *Isoodon obesulus*: Postures of Short-nosed Bandicoots in captivity; the animals figured were captured by Dr J. W. Kirsch at Roleystone, near Perth, W.A.

PLATE 26   *Perameles nasuta*: Postures of the Common Long-nosed Bandicoot in captivity; the animal illustrated was collected in the vicinity of Sydney by Dr Gordon Lyne.

PLATE 27   *Chaeropus ecaudatus*: Drawings of details of a preserved specimen of a Pig-footed Bandicoot from Central Australia in the Baldwin Spencer collection of the National Museum of Victoria; the specimen is preserved in alcohol.

PLATE 28   *Macrotis lagotis*: Two postures of a Dalgyte collected in Dampier Land near Broome, W.A.; the animal was sent to Perth by Mr John Tapper.

PLATE 29   *Dasyurus geoffroii*: A female Chuditch and her young from the south-west of W.A.; the animals were held captive at the Zoology Department, University of W.A., by Miss J. Arnold.

PLATE 30   *Sarcophilus harrisii*: Two Tasmanian Devils held captive at the Zoology Department, University of W.A.; the animals were supplied to the Western Australian Museum by the Animals and Birds Protection Board, Tasmania.

PLATE 31   *Phascogale tapoatafa*: Postures of a near-adult Wambenger from the south-west of W.A.; the animal was held captive by Mr L. McKenna; drawings based upon photographs with detail added from the captive animals.

PLATE 32   *Dasyuroides byrnei:* Postures of a Kowari from Sandringham Station, south-western Queensland, collected for Dr Patricia Woolley; the postures are from photographs by Dr J. W. Kirsch with detail added from the captive animal.

PLATE 33   *Dasycercus cristicauda:* Postures of a Mulgara from Kathleen Valley, near Wiluna, W.A., collected by Mr T. Moriarty and held captive by Mr E. Garratt.

PLATE 34   *Antechinus flavipes:* Postures of a Yellow-footed Antechinus in captivity; the animal was collected by Constable R. Taylor near Kalamunda, W.A.

PLATE 35   *Planigale* (close to *P. ingrami*): Postures of a pigmy antechinus in captivity; the animal was collected by Dr and Mrs E. Pianka near Laverton in central W.A. by digging a small pit into which it fell.

PLATE 36   *Sminthopsis murina:* Postures of a Common Dunnart in captivity; the animal was collected by Constable R. Taylor near Kalamunda, W.A.

PLATE 37   *Sminthopsis crassicaudata:* Postures of three captive Fat-tailed Dunnarts; the animals above and on the left were collected by Edwin Ride near Tambellup, in the South West of W.A.; the other was collected for Dr Gillian Godfrey (Mrs Crowcroft) on the Eyre Peninsula, S.A.

PLATE 38   *Antechinomys spenceri:* Four postures of a female Wuhl-wuhl caught by Mr A. S. Hill near Wiluna, W.A.; all but the central posture are based upon photography, with detail and texture added from the living captive.

PLATE 39   *Myrmecobius fasciatus:* Postures of a captive Numbat collected near Armadale, W.A.

PLATE 40   *Thylacinus cynocephalus:* Two postures of a Thylacine taken from a film in the possession of Sir Edward Hallstrom; the Thylacine was a captive at the Hobart Zoo in the early 'thirties; the rearing posture is from a sequence which records the reaction of the animal to a person rattling the bars of its cage; detail added from photographs of other Thylacines and from museum specimens.

PLATE 41   *Notoryctes typhlops:* Studies of various aspects of two preserved specimens of adult females of the Marsupial Mole from the Western Desert (near Kulgara, N.T., near Warburton Range, W.A.). They were collected by Mr W. Greenwood and Mr J. Carr and are preserved in alcohol in the Western Australian Museum.

PLATE 42   *Rattus fuscipes:* Two postures of a captive Southern Bush-rat from the South West of W.A.; the specimen was collected by Mr W. H. Butler at the Waychinicup River near Mount Many Peaks.

PLATE 43   *Hydromys chrysogaster:* A posture of a captive Water Rat from Barrow Island, W.A.; the background is from a photograph by Dr B. R. Wilson of its marine environment on the island; the foreground from the seashore at Cottesloe, W.A.; the animal was collected by Mr W. H. Butler.

PLATE 44   *Mesembriomys macrurus:* Postures of a captive Golden-backed Tree-rat; the animal was an inhabitant of the roof of a building in the pearl culture settlement at Kuri Bay in West Kimberley, W.A. and was captured by Mr K. Sheils; postures are taken from photographs with detail added from the living animal.

PLATE 45   *Leporillus conditor:* A drawing of Stick-nest Rats and their stick nest based upon photographs supplied by the South Australian Museum of a mounted group prepared by Mr J. Rau, who had kept the animals (which were from the Lake Eyre region) captive while they built their nest; he then stuffed the animals and placed them around it in appropriate postures; details are drawn from a spirit-preserved specimen, collected by Professor Wood Jones on Pearson Island, S.A., now in the National Museum of Victoria.

PLATE 46   *Notomys mitchellii:* Postures of a captive from the northern wheat belt of W.A.; the animal was collected by Mr W. H. Butler.

PLATE 47   *Zyzomys argurus:* Postures of a captive Common Rock-rat from Barrow Island, W.A.; the specimen was collected by Mr W. H. Butler.

PLATE 48   *Mastacomys fuscus:* Two views of an adult female from Kosciusko State Park; drawn with permission of Mr J. H. Calaby from photographs by Mr E. C. Slater published as Plate 1 by J. H. Calaby and D. J. Wimbush in *C.S.I.R.O. Wildlife Research*, 1964, vol. 9, pp. 123–33.

PLATE 49   *Pseudomys forresti:* Postures of a captive specimen of Forrest's Mouse from Thevenard Island, near Onslow, W.A.; the animal was captured by Mr and Mrs Peter Slater.

PLATE 50   *Melomys sp.:* Postures of two specimens of a captive mosaic-tailed rat from Port Essington, N.T., collected by the Division of Wildlife Research, C.S.I.R.O.; the species has not yet been identified with certainty.

PLATE 51   *Macroderma gigas:* A group of Ghost Bats, from a colony at Nullagine, W.A., the pale grey and white desert form; they were held captive by Mr Athol Douglas who collected them.

PLATE 52   *Nyctophilus timoriensis:* Postures of a captive Greater Long-eared Bat from the South West of W.A.; the flight postures are based upon photographs of the same bat flying in a closed room.

PLATE 53   *Rhinonicteris aurantius:* The head and nose-leaf of a newly-dead specimen of the Orange Horseshoe Bat from near Kununurra, W.A.; the specimen was collected by Mr K. Richards of the Western Australian Department of Agriculture.

PLATE 54   *Tadarida australis:* A captive White-striped Bat from near Katanning, W.A.; the animal was collected by Mr G. M. Doak.

PLATE 55   *Taphozous georgianus:* Postures of a captive Common Sheath-tailed Bat from Wilgie Mia in the Weld Range near Cue, W.A.; the specimen was collected by Mrs Helen Henderson.

PLATE 56   *Pipistrellus tasmaniensis:* Postures of a female Tasmanian Pipistrelle captured by Mr W. H. Butler near Pumphrey's Bridge between Pingelly and Narrogin in the South West of W.A.; the flight postures are based upon photographs of the same bat flying in a closed room.

PLATE 57   *Pteropus scapulatus:* Two postures of an adult Red Flying-fox in captivity at the South Perth Zoological Gardens; the animal originated from King Sound, Kimberley District, W.A., where it flew aboard a ship.

PLATE 58   *Syconycteris australis:* Two postures from photographs by Mr Stanley Breeden, of the Queensland Blossom Bat; the specimens were caught near Murwillumbah, N.S.W., by Dr John A. E. Nelson.

PLATE 59   *Canis familiaris:* The drawing of the Dingo is based upon photographs taken of a young male north of the Nullarbor, near Neale Junction, with detail added from animals captive at the South Perth Zoological Gardens.

PLATE 60   *Neophoca cinerea:* Two bull Australian Sea Lions; drawing based upon photographs taken at Carnac Island, near Fremantle, W.A.; and also from film taken at Dyers Island by Mr W. Gill, with detail added from a frozen specimen and a stuffed specimen in the Western Australian Museum.

PLATE 61   *Tachyglossus aculeatus:* Two postures of a captive Echidna which was collected near Two Peoples Bay in the South West of W.A., by Dr J. W. Kirsch and Mr A. Baynes.

PLATE 62   *Ornithorhynchus anatinus:* Postures of the Platypus based upon films and a number of photographs from various sources.

INDEX

Bold faced type is used to refer the reader to the summaries of names, distributions, habitats and external characters given at the end of each Group; italic type refers to entries in the list of scientific descriptions in Appendix II.

Ride, W   **D**   **L**
    A guide to the native mammals of Australia ₍by₎ W. D. L.
Ride. With drawings by Ella Fry. Melbourne, New York,
Oxford University Press, 1970.

    xiv, 249 p. illus., col. maps (on lining papers) 27 cm. 7.50
                                       ANL

    Bibliography : p. 213–235.

235179

    1. Mammals—Australia.   ɪ. Fry, Ella, illus. ɪɪ. Title.

    QL733.R5   599′.09′94   79–17396
    SBN 19-550252-3   MARC

    Library of Congress   70 ₍2₎

DATE DUE

GAYLORD

PRINTED IN U.S.A.